David W. Cartwright, Mary F. Bailey

Natural History of Western Wild Animals

and guide for hunters, trappers, and sportsmen - Vol. 1, Second Edition

David W. Cartwright, Mary F. Bailey

Natural History of Western Wild Animals
and guide for hunters, trappers, and sportsmen - Vol. 1, Second Edition

ISBN/EAN: 9783337122867

Printed in Europe, USA, Canada, Australia, Japan

Cover: Foto ©berggeist007 / pixelio.de

More available books at **www.hansebooks.com**

"DAVID'S" RETURN TO CAMP.

NATURAL HISTORY

OF

Western Wild Animals

AND GUIDE FOR

HUNTERS, TRAPPERS, AND SPORTSMEN;

EMBRACING

Observations on the Art of Hunting and Trapping; a description of the physical structure, homes, and habits of fur-bearing Animals and others of North America, with specific rules for their capture; also, narratives of personal adventure.

BY

DAVID W. CARTWRIGHT.

(SECOND EDITION.)

WRITTEN BY
MARY F. BAILEY, A. M.

TOLEDO, OHIO:
BLADE PRINTING & PAPER COMPANY.
1875.

Entered according to Act of Congress in the year 1873,

BY D. W. CARTWRIGHT AND M. F. BAILEY,

In the office of the Librarian of Congress, at Washington.

TO THE

MANY WORTHY AND COMPANIONABLE MEN

WITH WHOM I HAVE

HUNTED AND TRAVELED

THE FOLLOWING PAGES ARE INSCRIBED BY

"DAVID."

PREFACE.

For the past ten years Mr. Cartwright has been occasionally asked to put into book form something of his observations of the habits of many animals, of his knowledge of woodcraft, and some account of his life as a woodsman. He has been for several years engaged a considerable portion of his time, during the sporting seasons, in acting as guide for parties of pleasure seekers. Some of these sportsmen have also asked him to put into readable shape some of his travels in the out-door world.

He is not a professed scientist, nor does he claim to be in possession of the practiced use of the vocabularies of conventional technicalities; but his knowledge of the wild animals of this country is the practical knowledge of their homes and their habits, and the frequent personal scrutiny of their physical structure, which the naturalist needs and desires for the successful pursuance of his studies, and without something of which his knowledge is as a dictionary full of meaningless words.

He is also not a professional book maker, and he knows that it is only by practice that there comes any great degree of perfection in any art or trade. What he gives you, he puts upon the basis of an experience of forty years, and gives it with that assurance that he believes should come of practical knowledge, as opposed to any hypothetical and visionary trash. As to the matter of the subjects herein set forth, his endeavor is to meet the requests of his friends, believing that he understands them. As to the manner of their presentation, referring more specially to the narrative portions of the book, his design is to embody the truth in its deserved credibility. He has personally but little interest in works of fiction, and less sympathy for the story-teller as such. He would, therefore, be foolish to desire to present the many incredulities of travel and adventure, as is often done, even if it were not also true that in such an effort he would be like a man in deep water, who could find no way of escape. Therefore,

> Though this work be ragged,
> "Tattered and jagged,
> Rudely rain-beaten,
> Rusty, moth-eaten;
> If ye take well therewith,
> It hath in it some pith."

In the course of his life as a hunter, he has come in contact with many inexperienced, and many unsuccessful woodsmen. He hopes to benefit that class of readers, by placing within their reach rules that he has in every case thoroughly tested.

Since the author of this book claims for himself an incompetency to the task of putting it into shape, and the more exact wording of its pages, and has placed that part of the work into the hands of another, it is due to him to say that he labors under difficulties in presenting this book to the public. There is of necessity in many places a lack of that clearness of ideas, and of that sympathy that can come only from one actually experienced in such a knowledge and such a life as it is now the effort to make known.

Two of the articles, "The Speckled Trout," and "A Trip to Lake Superior," were furnished by friends of Mr. Cartwright. With these exceptions, the work has been done by means of notes taken by me at the dictation of the author. In filling out the description of the animals, "The American Cyclopedia," "Chambers' Encyclopedia," an English "Museum of Animated Nature," and "The American Beaver," by Lewis H. Morgan, have been used as references. All other books thus used are acknowledged in their respective places of reference. When the notes were filled out they were subjected to Mr. Cartwright for revision, and stand as in the print by his authority. Some of the illustrations he has secured from the publishers of Wood's Mammalia; all others have been prepared expressly for this work.

It is likewise due to the writer of these pages that it be said that there are difficulties lying in her way which must be faced, but cannot be overcome. It is as if you were obliged to tell for others what they know and feel, and say it as well as if you knew and felt it for yourself, and yet you do not.

One man could write books, if he knew how to write them; another could make books if he had something of which to make them. M. F. B.

Milton, Wis., Oct., 1875.

CONTENTS.

PART I.

THE HUNTER'S ART AND HIS GAME.

Chapter.		Page
I.	Hunting as a Business	3
II.	Useful Hints on Woodcraft	8
III.	The Deer	12
IV.	The Antelope	39
V.	The Rocky Mountain Goat	41
VI.	The American Bison	43
VII.	The Prairie-Dog	45
VIII.	The Woodchuck	47
IX.	The Weasel Family	49
	The Pine Marten	49
	The Mink	52
	The Fisher	54
	The Badger	56
	The Otter	58
X.	The Skunk	63
XI.	The Wolverine	65
XII.	The Bear Family	68
	The Black Bear	71
	The Grizzly Bear	75
	The Raccoon	78
XIII.	Felidæ, or Cat Tribe	80
	The Canada Lynx	82
	The Bay Lynx	85
	The Cougar	86
XIV.	The Fox	89
XV.	The Wolf	96

CHAPTER		PAGE
XVI.	THE OPOSSUM	100
XVII.	THE MUSKRAT	104
XVIII.	THE BEAVER	109
XIX.	SPECKLED TROUT	134

PART II.

NARRATIVES OF PERSONAL ADVENTURE.

I.	MY FIRST FOX HUNT	141
II.	MY FIRST BEAR HUNT	143
III.	HUNTING IN ALLEGANY CO., N. Y.	146
IV.	AN ADVENTURE WITH A WOLF	155
V.	HUNTING IN JEFFERSON CO., WIS.	158
VI.	A TRAMP TO CALIFORNIA IN 1852	165
VII.	HUNTING TRIPS IN N. W. WIS. AND IN MINN.	235
	A TRIP IN N. W. WIS.	235
	IN THE CHIPPEWA REGIONS IN WIS.	238
	IN THE WOODS IN N. W. WIS.	240
	ABOUT EAU CLAIRE	241
	IN EAU CLAIRE AND DALLAS COUNTIES	246
	TRAPPING IN MINN.	248
	IN THE COTTONWOOD COUNTRY	253
	AN ENCOUNTER WITH AN EAGLE	254
VIII.	A TRIP TO LAKE SUPERIOR	256
IX.	TRAPPING IN THE LAKE SUPERIOR REGIONS	271

ILLUSTRATIONS.

	Opposite Page
FRONTISPIECE	
VIRGINIAN DEER	17
ELK OR MOOSE	36
WAPITA OR ELK	37
WISH-TON-WISH OR PRAIRIE-DOG	45
PINE MARTEN	49
MINK	52
FISHER	54
BADGER	56
BLACK BEAR	71
LOG-TRAP FOR BEARS	75
CANADA LYNX	82
WILD CAT	85
COUGAR	86
RED FOX	92
VIRGINIAN OPOSSUM	100
MUSQUASH OR MUSKRAT	104
AMERICAN BEAVER	109
TROUT FISHING SCENE	134

PART I.

THE HUNTER'S ART AND HIS GAME.

I.

HUNTING AS A BUSINESS.

Before attempting any description of the various methods of capturing different animals, I wish to make some statements relative to hunting and trapping as a business. A vast number of people count the hunter's and trapper's occupation as dishonorable, fit only for roughs and idlers. The business, as a business, is not dishonorable. It cannot be unless it is dishonorable for one to buy, sell, or use for wearing apparel, or for any mechanical purpose, or for the food markets the furs and pelts of animals, the skins, bones, horns, hoofs, and sinews, and the palatable flesh of many of those same animals. As a point of fact, however, it is probably true that the odium which is cast upon the business, as such, comes from the general understanding of the people relative to the character of a hunter as a hunter.

The man who earns his living by his rifle and his steel trap, and who supplies the demands of the various markets, must of necessity work far back from the thickly settled portions of the country. The communications between him and his buyers is such that he spends but a small portion of his time in the so-called civilized world. There are comparatively few people who become personally acquainted with this wild man of the woods. There are in by far too many communities men who neglect their families and their business to hunt and fish, who care less for their homes than their hunts, their fish than their fun. There are also men in almost every town who frequently spend one day or several days in hunting and fishing, but who do it professedly for the fun. They return to their homes tired, and generally crippled in some way. They have hunted so near the homes of some of their neighbors that they generally find but little game, and, as a rule, have used more ammunition than their game would be worth in market. It is not, therefore, strange, owing to the easy way in which we dispose of so many questions which do not intimately concern us, that people, looking upon the two classes of hunters just referred to, count the real hunter unprincipled, as is the former class, and unsuccessful as is the latter. The hunter and trapper need not be a rough or an idler, and, indeed, if he is a successful business man he cannot be either. The rugged constitution necessitated by the exposures and hardships of such a life as his is the only sort of roughness which is in any way demanded by the business. A skillful hunter will find, in any gamy country, enough to do to keep himself busy.

Hunters are not without honor. There are rules of honor which hunters, as a class, consider themselves bound to respect, and which I have never known any one of them to violate. Respecting the right of territory trappers have long since decided that the first man upon a certain ground had a right to that ground so long as he should choose to hold it. If a trapper finds certain ground occupied, as he can determine by traps set, or by occupied camps, he leaves it. If, in going the rounds of his traps, a trapper finds that another man has set traps upon his ground he hangs them up, and passes on. When the man who has set those traps finds them thus removed, he at once understands that he is upon occupied ground. There are no territorial rights amongst hunters; but, if a hunter has started up game and it runs upon parties who may be hunting close by, and they find it coming to them already wounded, they may kill it and take one half of the meat, but the other half of it, and the whole of the hide belongs to the man who first shot it. He may follow up his wounded creature; but, if he finds it in the hands of a second party, he cannot and does not claim the whole of the flesh.

Does it pay to hunt? Yes. It can be made to pay well. It would not pay to spoil the good blacksmith to make a poor hunter; but, given a man of strength, of physical powers, courage, endurance, a close observation of the habits of the animals hunted, skill in the methods used for their capture, ability to live in the woods without getting lost, a good trapping and camping outfit, and with all these a liking for the business, and the result is monetary success. For the frontiersman, thus qualified, it is emphatically money in the pocket. There is every year a portion of the time when his farm work does not keep him busy. By proper management he can then earn several times as much as if he were to hire out by the day to do any ordinary kind of work. Speaking from experience, I believe that I can safely say that during those years when I lived upon a frontier home, I earned five dollars per day for every day that I hunted, which was certainly more than any neighbor of mine could pay me for working for him, even if he had had work to be done, and was every time five dollars more than I could have received for doing little or nothing upon my own place. It is also an advantage to a frontier settler to have a good knowledge of hunting and trapping, as he can thereby increase the bounty of his board, even if he never traffics in skins or flesh of wild animals. He also hastens the more civilized condition of his home surroundings, by thinning the ranks of the wild creatures about him.

A great many of the hunters are jolly, genial, free-hearted men, and when they have come out of the woods they gratify their generosity, and their fun-loving propensities by quickly spending all they have just earned, and are thereby forced to go back again into the woods. I would advise any young man who *cannot* find other employment, to go into the woods, and if he does not know how to hunt, to learn, rather than being out of business, to lounge about the loafer's resorts, and thus become a nuisance to a community and a bore to himself. I would not for any other reason advise one who is not specially adapted to the work, to attempt to follow it as a business. It is such a man's fault more than any other one's that every hunter must bear the reproaches so generally laid upon him.

I could never advise an intemperate man to become a hunter. Alcoholic spirits are not necessary to ensure safety against the exposures of life in the woods. A vigorous constitution which has never been damaged by the fitful fires of alcohol will of itself preserve an even temperature of healthy blood and spirits. The hand and the judgment of a drinker are unreliable, and thus unfit him for such work. The use of tobacco is no less than suicidal to every attempt at successful hunting and trapping. If one could catch game by driving it from him, no better scent could be used than that of tobacco. Knowing, as who does not, that the whole brute creation abominates the scent of tobacco, it is a mockery of all honest pretensions for a man to claim that he reaps the full benefit of his hard hunter's toil while he habituates himself to the use of tobacco by any means whatever. If he does not believe this, it must be because he has never made a fair trial of hunting with no tobacco about him. I have in a few cases been upon hunting grounds with men otherwise well adapted to the work, but who, by their tobacco-using habits have greatly reduced the success of the trip, and I have been obliged to abandon a field because some one of the company would persist in using tobacco.

II.

USEFUL HINTS ON WOODCRAFT.

A trap should always be set with one end pointing towards the probable approach of the animal for whose capture it is intended. If a trap is set in a hole, one spring should be in the hole, the other out of it; otherwise the jaw will be very likely to throw the leg out of the trap as the trap springs.

Never put bait on the pan of a trap. The bait should always be put either above or beyond the trap; its position is to be determined by the kind of game for which it is set. Scarcely any animal can be caught by the nose, as it should be if the pan were the place for the bait; scarcely any animal would be caught by the nose if it could be; for the attention of the animal would be directed towards the trap. The design is that the animal should step upon the trap while trying to secure its bait, and if the trap is not baited and is concealed, no attention is drawn towards it. The animal then knows nothing of the trap until it is sprung.

It is always better to cover a trap when set, if it can be done, even for animals of aquatic habits. When a trap is set in water for a coon it does not need to be concealed; otherwise the rule holds good. Mud, or sand, or a light covering of earth, grasses, or leaves, or small twigs, or other light stuff that may be easily picked up are the materials used for concealing traps.

For capturing heavy animals a clog or weight should be used. I would, however, except the deer; for they will not go off far carrying a trap with them, and they are apt to bruise themselves worse if the trap is clogged. The clog is usually a pole, over the larger end of which the ring of the trap-chain is slipped and then fastened by a wedge.

A spring-pole is a pole so fixed as to spring when displaced by the motions of an animal in its attempts to extricate itself from a trap, and to fling the animal into the air, and there holding it to keep it safe from the clutches of any beast of prey, until the trapper shall secure his game.

The sliding-pole is used to secure aquatic animals from molestation, by drowning them as soon as the trap springs upon them. As soon as an aquatic, or an amphibious animal is caught in a trap it plunges for deep water. For small game the weight of the trap is sufficient to drown them, but for larger animals a sliding pole is used. A pole about twelve feet long is taken and is stripped of all its branches, and a nail is driven into it on its upper side, in a slanting direction, near the smaller end of the pole, the head of the nail pointing towards that end. The pole is then placed near the trap set, which is very near the bank, the ring of the trap chain is slipped upon it. The larger end of the pole is fastened by a hook put over it and driven into the ground, but the pole is to be secured high enough from the ground for the ring to have a chance to traverse freely down the length of the pole; the smaller end of the pole is fastened in the bed of the stream. When the animal springs the trap and makes its lunge for deep water, the chain ring slips down upon the pole and carries with it the trap and the animal, but once past the nail it is held from returning: the animal must therefore drown.

The Newhouse *steel trap* is the best one in use. It is perfectly adapted to all the purposes for which it is designed, except No. 4 for deer. I consider No. 4 worthless for deer catching. A trap to catch a large buck should be nearly as strong as that required for a bear. I therefore use No. 5.

The number of traps required for a good outfit depends upon the kind of game to be trapped and the country in which one traps. If a man is going into the woods where there are no roads, and where he must carry his baggage upon his back, he will naturally limit his supplies to the actual necessities of the case. To fill this demand I would advise two dozen traps of Nos. 3 and 4, or large ones, and thirty-five small ones. If he can by boat reach the heart of his trapping ground, he can take as many as he pleases to carry; but it is never advisable to take more than the trapper can tend. From seventy-five to one hundred is as many as two persons can well tend under ordinary circumstances.

It is preferable to have a trapping ground where one can travel mainly by water. It is easier, is pleasanter, and is more profitable. Light but serviceable boats may be easily made by any one of ordinary ingenuity and knowledge of boat building.

For a camping outfit, I would advise that *nothing* be taken that can well be dispensed with. Inexperienced hunters almost always cumber themselves with needless articles. Two heavy, double blankets are necessary for two persons. A small pail for tea or coffee, a small camp kettle, a frying-pan and a tin cup should be carried, and one knife and fork and two tin plates for each person.

For food-stuffs I take flour, corn meal, beans, a little pork, some butter and sugar, and unless going very far into the woods I carry potatoes. *A pocket knife, a good compass, matches, and a hatchet are indispensible to a hunter's outfit.* No man should ever go out from camp without them.

III.

THE DEER.

The deer family has been variously classified by naturalists, some giving to it several genera, others regarding its groups not sufficiently marked to give to them generic character. Most writers base their classifications upon the horns. The horns, however, are not found upon all species, nor at all seasons of the year, and they also assume a different aspect at different ages of the animal, and similar horns, in some instances, grow upon species otherwise distinct. Some have therefore classified them with respect to the kind of hair which forms the fur, to the form and extent of the muzzle, and to the position and presence of glands on the hind legs. Again, they are classified as follows: First, those of snowy regions, having a broad muzzle, hairy palmated horns, a short tail, fawns not spotted; Second, those of temperate regions with a tapering muzzle, ending in a bald muffle, fawns and sometimes adults spotted.

The deer family is represented in almost every region of the globe, Australia and central and southern Africa being the only exceptions to this universality of distribution. Hills of moderate elevations, wide plains, and forests are the localities to which these fleet-limbed creatures give the preference. The most of them herd together in troops; some few live singly. They use their powerful horns for weapons of defense, and sometimes of offense; but in general they trust to flight for safety.

They vary in size from the Pigmy Musk Deer of the Asiatic islands, which is not larger than a hare and weighs only five or six pounds, to the gigantic Moose which attains the height of seven feet at the shoulders, and weighs twelve hundred pounds. They vary in color in the different species, and also in the same species at different seasons of the year. They are timid creatures. In form they are light and elegant, and combine much compactness and strength, with slenderness of limb and velocity of movements. They furnish food and clothing, and are used as beasts of draft for many northern nations.

The skeleton is constructed for lightness and rapid, springing motions; and the spines of the dorsal vertebra are long and strong, thus suiting them for the origin of the thick ligaments necessary to support the ponderous head. The cavity of the skull is small, being in conformity with the limited intelligence of the group. The neck of a deer is long, thus adapting the animal for grazing; its head is small, and this it carries high. The eyes are large and full; the pupils are elongated. In most of the species there is below each eye a sac or fold of the skin, varying in size, called suborbital or lachrymal sinuses or tear-pits. Their use is not fully understood; but they doubtless serve some important purpose in the animal economy. They secrete a peculiar unctious fluid, said by some to be the most profuse during the rutting season. The ears are large. The tongue is soft. There are eight cutting teeth in the lower jaw, but none in the upper jaw. The males have usually two short canine teeth in the upper jaw, but neither sex has any in the lower. The prae molars are three in number, and there are three true molars on each side in each jaw. The deer is clothed with hair, which is longer and thicker in cold countries than in warm. The feet end in two toes, each with its sharp hoof resembling a single hoof which has been cleft. Behind and above there are two small rudimentary toes or hoofs. The two metacarpal and metatarsal bones are united into a single bone.

The deer is a ruminating animal. It is distinguished from all other ruminants by branching antlers or horns, which, in the most of the species, exist in the males only. These horns are solid, and are lost and renewed annually. They increase in size and breadth of palmation until the animal has become old, when the horn diminishes in size at each annual renewal. The reproduction of the horn is the most active during the rutting season. It is not until spring, or the beginning of the second year that the first pair of horns begins to make an appearance. When these do begin to grow, it is sometimes the case that one is in advance of the other. I believe it to be an erroneous idea, which so many have, that the deer grows a pair of horns when one year old, and that for every additional year there is always an additional prong to each horn. The growth of the horns depends entirely upon the condition of the deer. Sometimes prongs will appear upon the horns when the animal is but two years old, and the next year there may be none added. Sometimes there are three prongs on one horn more than on the other. The process by which the horns are developed, die and are shed is a very curious one, and is described in the "Museum of Animated Nature," as follows: "The skin enveloping the peduncles swells, its arteries enlarge, tides of blood rush to the head, and the whole system experiences a fresh stimulus. The antlers are now budding, for on the top of these footstalks the arteries are now depositing layers of osseous matter, particle by particle, with great rapidity; as they increase the skin increases in an equal ratio, still covering the budding antlers, and continues so to do, until they have acquired their due development and solidity. This skin is a tissue of blood-vessels, and the courses of the large arteries from the head to the end of the antlers are imprinted on the latter in long furrows, which are never obliterated. In ordinary language the skin investing the antlers is termed velvet, being covered with a fine pile of close, short hair. Suppose, then, the antlers of the young deer now duly grown, and still invested with this vascular tissue; but the process is not yet complete. While this tender velvet remains, the deer can make no use of his newly-acquired weapons, which are destined to bear the brunt of many a conflict with his compeers; it must therefore be removed, but without giving a sudden check to the current of blood rolling through this extent of skin, lest by directing the tide to the brain, or some internal organ, death be the result. The process, then, is this:—as soon as the antlers are complete, according to the age of the individual, the arteries at their base, where they join the permanent foot-stalk, always covered with skin, begin to deposit around it a burr, or rough ring of bone, with notches through which the great arteries still pass. Gradually, however, the diameter of these openings is contracted by the deposition of additional matter, till at length the great arteries are compressed as by a ligature, and the circulation is effectually stopped. The velvet now dies for want of vital fluid; it shrivels, dries, and peels off in shreds, the animal assisting in getting rid of it by rubbing his antlers against the trees. They are now firm, hard, and white; the stag bears them proudly, and brandishes them in defiance of his rivals. From the burr upwards, these antlers are now no longer part and parcel of the system; they are extraneous, and held only by their mechanical continuity with the foot-stalk on which they were placed; hence their deciduous character, for it is a vital law that the system shall throw off all parts no longer intrinsically entering into the integrity of the whole. An absorptive process soon begins to take place just beneath the burr, removing particle after particle, till at length the antlers are separated and fall by their own weight, or by the slightest touch, leaving the living end of the foot-stalk exposed and slightly bleeding. This is immediately covered with a pellicle of skin, which soon thickens, and all is well. The return of spring brings with it a renewal of the whole process with renewed energy, and a finer pair of antlers branches forth."

The origin of the horns is called the burr, the main shaft the beam, and the branches antlers; when near the head they are called brow antlers. If the brow antlers grow forward over the face, they are called by hunters the "looker prongs"; the termination of the beam is called the perch, and the small processes are called snags and prickets.

The horns begin to grow in April, and attain their full size in August. They are shed, according to the condition of the animal, any time from December to February. The color of the horns is a pale red when the antlers first branch through the velvety skin, and by a gradual change it comes to be by October a yellowish brown, and by the time they are again ready to fall off are very light colored, sometimes said to be white.

The species in which American trappers are most practically interested are the Virginian deer and the Black Tailed deer.

THE VIRGINIAN DEER.

The graceful creature known as the Virginian deer is the most useful of the wild game of North America; it is valuable to the red man and the white, to professional hunters and trappers, and an unfailing source of delight to the sportsman. It is also interesting to the naturalist, because of its physical beauty, and the peculiarity of its habits. The flesh of the deer is counted by many a very desirable article of food. The Indians eat it in preference to almost any other meat. Of its horns are made handles of various kinds of cutlery; of its skin clothing is made; its sinews form the bow-strings and the snow-shoe netting of the North American Indian. The Virginian Deer is found in nearly all of the states of the Union east of the Rocky Mountains. Deer are gregarious, though frequently found alone. The average herd is from four to seven—I have seen thirty-two in a herd.

It is not true, as is generally supposed, that deer are fond of grass. They do not like to eat timothy grass, and will not if other food can be found. Their food in summer consists of berries, nuts, roots, twigs, and persimmons; they are especially fond of buckwheat when it is just peeping through the ground, and is very tender. They delight in going over recently burned districts, and feeding on the tender weeds, and on young raspberry shoots. Hunters take advantage of this habit when they expect in a few weeks to hunt deer. They like what is tender and juicy. In the summer they frequent lakes and rivers to feed upon the water plants. They are very fond of the pond lily, eating both blossoms and leaf; they are also fond of the mosses that grow upon stones in the water. They find under the water a plant which grows about a foot high, and for which they will throw the head far under, and if need be, nearly immerse the body to get it. Hunters call it deer cabbage. They are very fond of fruits of almost every kind. They do not crop their food by the mouthful, as does the cow, but select here and there a tender leaf or twig. They frequently visit the pioneer's clearing, appropriating his wheat, corn, oats, turnips, and cabbages; all of which they like to eat. In the autumn they feed largely upon acorns and other nuts. In the winter they retire to thick groves and woods, and to swamps, but do not go into the densest region of the forest, nor to the thickest of the swamps. They resort to these places to find the cedars which grow along the edges of the swamps. They are very fond of the white or yellow cedar, which grows in such regions. Moss, barks and browse are their chief support during the winter.

The Virginian Deer may be known by the peculiar shape of its horns, which in the adult male are of moderate size, bent boldly backwards, then suddenly hooked forwards. The average weight of the horns is four pounds, that of the animal, at live weight, is two hundred pounds.

Cast horns are said to be rarely found. It is true that they are rarely found in countries where the deer are themselves seldom found; but in regions where deer abound, their horns are often seen lying upon the ground. The animal does not cover its cast horns; but leaves them and pays no attention to them whatever. They may, of course, be at times covered; but, if so, not by any effort of the deer to secrete them, but by a perfectly natural process, as the falling of leaves, or of trees, or in the northern regions by deep snows. I have many times been in the woods where deer were plenty, and their cast horns were scattered about in great numbers, there being in all probability not a single pair of horns secreted by any animal. They are eaten by the flying squirrel, and a mouse, called by the hunters the buck mouse. I have seen deer's beds, at the time of casting horns, where the antlers have evidently fallen, the one to the right and the other to the left of the animals as they had lain in those beds, and were undisturbed by them. I have shot deer having but one horn, the other being but just cast, and untouched by its owner.

As to the time of casting horns, the Virginian Deer forms no exception to the rule given for the family at large, unless it be that they sometimes shed them earlier in December than many species, and later in February. Sometimes they cast them even later than in February. In cold countries, and in unusually cold seasons they cast them very early in December, while in the same regions in open winters much later in the season. This is true of deer in the Lake Superior regions. In 1849 I killed a deer on the morning of the 12th of April, in Wisconsin, and both horns were solid on the head. This was of course an exceptional case.

Sometimes the horns remain upon the head until they have become so loose as to fall by their own weight; sometimes they may rub them off with their feet; or they may be pulled off by the brushwood, or pushed off by branches or trunks of trees when running through thickets; or they may be tossed off by a sudden and violent shake of the head.

It is said that bucks often lock horns, and that it is sometimes literally a "dead lock." Mr. Wright speaks, in the American Naturalist, of killing a pair of bucks so firmly united that they would have died of hunger, had he not killed them. Wood cites an instance in which three pairs of horns were interlocked, the skulls and skeletons attesting the deadly nature of the combat.

The hair is shed twice a year; the color of the animal is, therefore, different at different seasons of the year, and so continuous is the change of color that an observant hunter can tell within a month of the time when the deer has been killed, or has died. The summer coat is called red, it is about the color of a red cow. The bucks are of a brighter color than the does. The fall color is called blue; it is about a slate color. The winter coat is whitish. In June the deer wears its reddish coat, by August it has changed.

As the hair grows long for winter wear, the skin grows thin. In January, February, March, and April the skin is almost worthless. In the fall, when the deer is in its best condition, the reverse is true, the skin is the most serviceable, but the hair is not good. The hair is hollow, soft, and of great service in commerce.

The color of the deer on the lower parts, from the chin to the end of the tail, is whitish. The young, until four months old, are of a bright reddish brown, with irregular, longitudinal white spots; they afterwards resemble the old ones.

They measure five feet and four inches from the nose to the root of the tail. The length of the tail, including the hair, is twelve inches, the bones being only six inches long. The ear is five and a half inches high. The deer of two years has two teeth, of three years four teeth, of four years six, and of five years a full mouth. The age of the animal may be quite accurately determined by its teeth. By the time it is five years old it has shed its fawn teeth. Its permanent teeth are short and quite broad; as it grows old its teeth become quite pointed. It is by the number and shape of the teeth that the age may be determined. The average age of the deer is twelve years.

The deer has large, lustrous, nearly black eyes. The eyes are so full that the animal is readily apprised of danger coming far to the rear, and unless an animal or person directly behind it comes with the fewest possible motions will be detected. The senses of hearing, seeing, and smelling are very acute; but the last one named is the only one that can, of itself, insure the animal safety. The deer is exceedingly timid, and persecution has increased its timidity. It soon gets acquainted with the voices and general appearance of its pursuers, and becomes doubly cautious of them. It will run sooner from the voice of a dog than a man. It will run from one Indian sooner than from several white men, though in point of fact the white man is the better hunter. The Indian will run as fast as possible upon the deer's track, driving it from him at every step; while the white man, with more skillful management, will kill or catch the deer. The deer has for so long been a shot for the hunter's rifle, the prize for the chase, and the prey of many animals, that it is easily startled by any of the thousand noises that wake the silent woods. The sense of vision is at once called into play, and in most cases if the motion is inconsiderable it will recover confidence when apprised of the cause. The form and color of a strange object have but little effect in frightening a deer; it is motion that draws its attention.

The deer may approach very near a person, and, so long as the man is still, he fears no danger. A man may also approach very near to a deer. One day when accompanied by a fellow hunter I saw a deer at some distance from us, which my companion thought we could not approach near enough to shoot. The deer stood on one side of a marsh; we were on the opposite side, and there was nothing between us that by any motion could possibly frighten it. It was at the time feeding. When a deer feeds it does not stand still, as for instance the cow, but keeps on walking, nibbling and biting here a little, and beyond a little. When feeding, before it steps along it will whisk the tail about, look around to see if all is right and it is safe. If all is well, it will browse again; but if not, and it is at all frightened, it will bristle up the tail and raise it erect. Since the tail is white on the under side, it may be seen in this position at some distance. By taking advantage of these facts, so well known to hunters, I came within thirty rods of the deer and killed it.

When the deer looks up, if it sees a man or any frightful object, it looks at it steadily; if the man moves, the deer will very soon run; but if he remains still, the deer will look to the right and to the left, drop its head and feed again, look up again at the man, to the right and the left, and again drop its head. After proceeding in this manner for some time it will lose its fear. The deer keeps its head down but a very short time, probably not more than half a minute; the hunter can, therefore, approach the animal only by slow degrees, as he must step while the deer's head is down, and remain motionless while the head is up. When he has come sufficiently near to it to fire, all his preparation for firing must be as gradually performed as was his advancing upon the animal. The approach upon a single deer is thus comparatively easy. When two are together feeding it cannot be done so easily; for they will not, unless by chance, and then only for a very short time feed at the same time, and indeed such chances seldom become verified in fact. The difficulty increases with every deer in the herd. When several are together, it is many times useless to attempt to get within shot of them, but better to seek a smaller herd. Hunters upon the prairie pursue this method. In the woods the hunter is secreted by trees or bushes.

It is worthy of note to the hunter and to the sportsman that the deer will be more easily caught at feeding times than when lying down, as they are in the latter case always on the watch. In the morning, at noon, and at night deer are feeding. The early morning is the best time of day for hunting them. They feed until the sun is about an hour high, then lie down until about eleven o'clock; after this they feed for an hour or two, and again lying down they remain until near sunset, when they get up and feed more or less, but probably do not lie down much during the night, but play and race about.

I said that deer will sometimes approach very near a person. One day Mr. Streeter and I, who were at the time hunting together, were sitting one on either side of the trunk of a fallen tree. A little poplar stood close beside us. We saw a deer approaching, and as it evidently had not yet seen us, we decided we would not touch it, but by remaining motionless see how near to us it would come. We remained quiet, and the deer had come close to us and was feeding by the branches of the fallen tree before it saw us. It then snorted, or "blowed," as the hunters call it, seemed frightened, looked me steadily in the face, but soon began to feed again, at every step nearing us. Soon it cropped some leaves from the little tree close by my side. We began to whisper to each other, and then to talk to the deer. We told it to come on, that we were pledged not to hurt it. It did not come, but jumped back a few feet, and as soon as its senses of vision and hearing had satisfied it of the cause of its fear, our little game was played, and we let the creature go as it pleased.

The deer's sense of smell is so very acute that if it smells its danger it is not so easily bewildered. No very near approach can be made upon the deer, nor will be made by the deer if the wind blows from the object of fright upon it. Hunters must therefore advance "up" or "against the wind."

Deer emit but few sounds that may be termed vocal. One sound is quite like that made by a calf when frightened. The fawn, when caught, bleats like a lamb under like circumstances, and the grown deer sometimes cries out in much the same fashion. Another sound the hunters call a "blow." It is a sort of snort or explosive whistle, but the emission of air is from the nostrils. When the deer snorts, it throws one of its fore feet forward and stamps with it upon the ground; just before it snorts it throws its ears back, but as it snorts, throws them forwards. The doe will always bawl when shot. The buck does not; neither does the buck blow as much nor as often as the doe. The doe has a fine, low, and pleasant voice or call for its young. This call is responded to by the fawn. This response is something like the bleating of a little lamb; but the sound is continued longer. It may be well to say here that though it is stated by reliable authorities that the terms Buck, Doe, and Fawn are incorrectly applied to the Virginian Deer, the Black Tailed Deer, and the American Elk or Moose, and should be replaced by the terms Stag, Hind, and Calf, that hunters throughout the countries where these species are found invariably call the male deer the Buck, the female the Doe, and the young deer the Fawn.

It is said that the old bucks consort together the most of the year, and that the does and young bucks go in herds by themselves. In the fall the males go singly, though they meet at their "pawing places," where they root up the dirt, bark trees, paw up the snow, and engage in fierce contests. It is rarely the case that two old bucks are found together when feeding or lying down, especially in the fall. The fawn does not, as some have said, often follow a man who is on horse back. If it has been carried by a man, and he puts it down, for a time it will follow him.

The leader of a deer herd is not necessarily the finest, the largest, or fattest one, and may not be a male deer. A female deer is quite as apt to be the leader of a herd, and the largest buck to be the rearmost of the herd. When a herd is following its leader, it takes the single file, and so exactly do they follow each other that a person observing the track could scarcely tell if there had been more than a single deer over it. A fawn is sometimes the leader of a herd. On one occasion I came suddenly upon two bucks and a fawn, the bucks being led by the fawn. As is always advisable when a hunter fires into a herd, I fired at the leader, this time the better creature, and killed it. I then shot the first buck, broke its back but did not kill it. The second buck had by this time gone off, and stood looking at me from behind an old log. I fired at it, and knocked out all the teeth on one side; the deer then started up, made a circuit about me, and came up in front of me and about twelve rods from me. I killed it and then went to my wounded deer and killed it. The five shots were fired within a very few minutes of each other, and I was the possessor of an unexpected prize.

The doe bears her young about the middle of May, or in June, the first, when she is two years old. She retires from the male deer, and carefully watches her fawns, though much of the time at a distance, probably because of her anxiety that nothing shall detect their hiding place. She secretes them so effectually that it is difficult to find them. About the last of July, when they are able to run about, as fast as she does, she brings them out from their retreats. For about six weeks they run with the doe, after which, until November or December, the doe hides from the fawns. She then returns to them, and through the winter they run about with or without each other. The doe generally has two fawns at a birth, and she rarely loses one. Probably one out of every twenty-five raises young every year.

The deer is not found in so great numbers as formerly, having been diminished by the hunter's rifle, and like the Indian and many wild beasts of forest and plain, driven before advancing civilization. Stringent laws are in force in some countries for the protection of these beautiful creatures. But nature, in all of her sources and resources, is still rich. She furnishes a system of compensations by which a weak spot has somewhere a support; a defenseless creature has still some protection. No dog, or other animal can scent a fawn, thus, while otherwise defenseless, it matures without molestation, and when mature, stands its chances with its fellows to escape all harm.

The males are prime from August to November. The does are at this time lean and in poor condition. The bucks are then both careless and fearless, and become an easy prey to the hunter: the doe, driven by hunger, is intent on feeding, and generally before midwinter comes into good condition. The bucks, in turn, have become lean, and the skin and flesh are worthless, the former emphatically good for nothing, and the latter having acquired a rank taste, is inviting to no man's palate, unless he be driven by extreme hunger.

Deer delight in immersing themselves in water during the hot weather, to rid themselves of such persecutors as the fly and the mosquito. They are excellent swimmers. Their bodies are deeply submerged, and they swim so rapidly that nothing but an Indian canoe can easily overtake them. They are enduring swimmers, and have often been seen crossing broad rivers, and swimming a distance of two miles.

About the 20th of August the deer which are in the northern regions begin to wend their way southward, going in small herds. They start early in the morning and travel until about noon, when they feed, lie down, and do not move on until the following morning. Later in the migratory season, the herds are larger and the marches longer. Those going last are again in small herds, but they will travel at the rate of one hundred miles a day, and will scarcely stop for food or sleep. A similar course is pursued in May, when the deer return northward. They may be caught when going either to the north or the south by building fences or other barricades across their paths. They are to be built in the shape of the letter X, and a watchful hunter can then easily capture them. They are the most easily captured, and may be taken in the greatest numbers at their migratory seasons; but what is true in very many instances is true in this case,—what costs the least, either in money or effort, is worth the least. It is not always advisable to catch whole herds of deer because they may be so caught.

In the winter, when the snow is deep, they make "yards" and paths by tramping down the snow. From these paths they reach from one side to the other for their food. They collect in large herds in their yards, and these are enlarged from time to time as the deer need more room for browsing. Wolves and panthers are here their most formidable foes, unless we except the hunter. Panthers, or more properly cougars, steal upon them on the sly, when in these yards. They crawl close to them, and spring suddenly upon them, or they watch them from some cliff or tree, and from these places pounce upon them after the manner of a wolf coming upon a fold. Wolves sometimes pursue a single deer with the "long chase." In the summer deer will avoid wolves by springing into the water.

Deer may be easily domesticated, but it is quite impossible to keep them tame. They will remain for months, and sometimes for a year or two about their new home, but are almost always sure to get back to their native haunts. Even while they may be supposed to be thoroughly domesticated, they do not lose their native shyness. They are the first to notice danger coming to the enclosures of a farm or yard. Were they not such troublesome pets they would be exceedingly desirable ones.

There are several methods used by men for catching deer. That most in vogue with sportsmen is known as the chase. For sportsmen this, will answer; as the object desired is gained in the exciting pleasure which the chase furnishes. It is not, however, a desirable method for the practical woodsman, though practiced by many of them. The hunter and the hound are not the only excited ones in the party. The excitement, which is the most intense with the deer, affects its whole physical structure to that extent that it spoils the flesh for food. The legs become immediately stiff as soon as the animal is chased down. The meat is dry, hard, and black, and the surface of the flesh has a frothy appearance as the hide is torn off.

Fire hunting is the method used most successfully in the summer. At this time deer resort to the water side, and the hunter prepared with boat, gun, and lamp may be quite sure of his game. The light is set on the bow of the boat, so that it will shine on the forward sight of the gun, and at the same time conceal the hunter by its glare. To furnish him more certain protection, there must be a board, measuring about eight inches in height, put between the hunter and the lamp. The front side of the board must be blackened. A very bright light attracts the eye of the deer. When the hunter is hidden from the sight of the deer, he may see the animal without being in return observed by it, and may, by proper steering, come within a very few feet of it, sometimes almost within reaching distance. The glare of the light being so great as to cover in its light the boat and the boatman, the deer, in its curiosity to see what it is, frequently advances to meet its enemy; the two thus advancing upon each other may often come within six feet of each other, and I have known hunters to run their boats so close to deer as to hit them. For this method of hunting the boat must be so steered that the bow shall point directly to the animal; else if the blackened board should be turned edgewise towards the deer, and certainly if turned so far as to hide any of the light, the deer will at once see the man, and the man is then very likely not to see it again. The oars must be muffled, and there must be no noise in the boat. By the reflected gleams from the eyes of the deer, the hunter is enabled to get a very good aim at it. Many are taken in this method in the early autumn. Another method of fire hunting calls into use the police lantern, which is put upon the hunter's cap. This lantern may be used in the water or fire-hunting just described, or it may be used on the land, if the hunter is where he can walk quietly.

The steel trap, as some affirm, is not much used for taking deer. That is a mistake, and more, the steel trap is used very effectively. Some also say that sportsmen regard the use of it for deer catching as barbarous. In point of fact it cannot be less barbarous than the sportsman's dog trap. In the latter case the long chase with the blood-hounds is made use of with the object and professed object of furnishing amusement or sport, as the name implies, to the sportsman. The practical deer hunter makes use of the steel trap; because it is an effectual means of its capture; because it is an easy method of capture, and is harmless to the flesh, the skin, or hair of the animal. It also subjects the animal to suffering of less duration than the chase; for the hunter, a part of whose business it is to watch his traps, takes the animal from its close quarters before it has been there long. Steel traps for deer catching must be large and stout, and must strike very high on the leg. They must be stout enough to break the bone, else it will not prevent their escape. They should be strong enough for a bear. The trap should be placed in the path of the deer where it crosses a stream, or enters a lake, and should be set under water, and in some way concealed by other covering than the water. If it is as heavy as it ought to be, it should not be fastened, or even clogged: the violence of the animal is so great when caught that unless it can drag the trap away with it for a little distance, or satisfy its nervousness by its desperate plunges, it will break loose.

The steel trap is of great service in what hunters call a "noisy time," a dry time, when the leaves, grass, or brush are so dry as to rustle loudly under the feet, and thus prevent a near approach.

The trap may also be set in their paths in the snow, and covered lightly by it. It should be set at some place in the path where the deer is to step over a stick or log. If there are no such obstacles in the way, put something there. Fell a maple or bass-wood tree, and let it lie across the track, or put there an old hemlock limb which has moss on it, that it may detract attention and cover the scent. I sometimes set a browse tree, maple, hemlock, or bass-wood as the country may furnish, and conceal the trap under the snow, close to the browse. The deer's sense of smelling is very acute: the trees just mentioned furnish the deer's favorite browse: they soon scent it, and approach it, and in most cases are soon entrapped.

Another method of deer catching we call the "still hunt." We then take them by following up on their tracks, or by watching their run-ways. When hunting by the still hunt the man should not follow precisely in the tracks of the deer, but follow along that track, going sometimes on one side of it and sometimes on the other, and frequently crossing it. Many hunters are unsuccessful in hunting deer by this method, because they follow exactly in the foot tracks of the animal, and if they do not at the time capture their prize, they lessen the chance for another day, as the deer is frightened, and takes a new track. Let the hunter, when he finds the deer's track, keep in sight of it, but not upon it, taking a circuitous, or better, a wave-like course along the track, and frequently crossing it. Let him now, though cautiously, and as if ignorant of the deer's nearness, take a circuitous route about the deer. The latter has, doubtless, by the acuteness of its vision, seen the man before he has seen it, and as a deer soon loses its restless fear when apprised of the cause of its fright, and will stand and watch the man, he may come about it in narrow and still narrower circles, until he shall come within shooting distance of it. The still hunt may be effectively used when the hunter is on horseback. This last method is, in my opinion, the best one. The manner of following the track shall be the same as that just described. If the hunter is so fortunate as to come upon a herd instead of a single deer, let him, having marked the leader, which he can readily do by watching the maneuvers of the herd, fire upon it. He must, if possible, hit the animal too far back upon the body to kill it. If it is too badly wounded to prevent its running off, the rest of the herd will not leave it, but will soon gather close about it. The hunter may then pick them off, one at a time, and when all are killed he may then kill the wounded creature. Sometimes when deer are very badly frightened they will not run, but will stop. It sometimes works well to put a bell on the horse's neck, as the deer can hear the bell before they can see the approaching horse, and will lose their fear of its sound before the hunter is in shooting distance of it.

When a hunter starts up a deer and it jumps from fright, and is about to run off, it will frequently pay him to give a short, shrill whistle, or a sharp crack of his gun, to arrest the attention of the animal. The deer will almost always stop for a moment when it hears such a noise. A sharp, quick noise of almost any kind, if loud enough, will arrest its attention. The skillful hunter may then, by cautious maneuvers, secure his game.

Deer will sometimes feed in the evening during the fall when they can get fall wheat, and rye, and they may then be shot by moonlight. To do this, a screen of some sort must be found, by which a gun may be set, and so sighted as to strike an object which may be ten or twelve rods distant, at a height of twenty-seven inches from the ground: behind the screen the hunter watches the approach of his game, and if it comes between him and his mark, his only care then is to fire his gun while he keeps his eye upon the animal. The shot is almost always sure, as it will hit a large deer, but rather low, and a small one, but rather high. This method is also a good one to use when deer are feeding or traveling in the woods.

Most people who know anything of the deer know what is meant by a deer lick, and know that the licks are counted favorite places for deer catching. In many cases there are natural licks. Where the water is soft deer will work at them well; but where it is hard they will not. They do not always prefer to drink the salt water, but very often choose to lick the salt from stones where it has crystallized. When artificial deer licks are made by the hunter, he should throw the salt in low, wet places, as the deer, when suspicious of coming danger, does not look to high grounds, but almost always to the lowest spots. The hunter, by taking advantage of this fact, can approach his game with much less danger of being seen. Sometimes salt may be put into old logs or hollow stumps. In selecting places for deer licks, the hunter should, if possible, be able to approach the deer from the higher ground, and when the wind comes from the animal upon him, that it may not scent him.

In cloudy or rainy weather deer will come out to feed at all times of day. In clear weather they will feed early in the morning, and again near night. Times and seasons are, then, to be observed in the watching of deer licks.

Amateurs and sportsmen frequently overlook deer, because they look too high or too far away for them. They may frequently be found behind a bush or log, either standing or lying down. If there is a fallen tree near you, look closely in its top, or if there is a knoll near by which is covered with brushwood, look sharp but low for them. Many inexperienced hunters look only for the tracks of the deer, as if by that means alone they could find them. This course is emphatically unadvisable, for when a man is once following a deer in its track, there is no more certain way of frightening it, and of keeping it beyond his reach. Deer are very apt to make some short turn just before lying down, and they are very likely to lie down upon some elevated spot where they can watch their back track. If, then, a hunter will *watch a track, but not walk upon it*, he can readily find his game when lying near its trail.

Such hunters are likewise apt to travel too fast, thus needlessly attracting the attention of the animal, and decreasing their chance of getting their game. Every motion must be made cautiously, quietly, and often very slowly. They are also too apt to think it advisable to kill a deer whenever they are within shot of one. Great precaution should be used by the hunter in choosing his game where deer are plenty and may be taken in great numbers, especially if he would make his craft pecuniarily successful. Furthermore, it is as essential to the monetary success of the practital woodsman that he have some knowledge of the relations of the supply and demand in the fur and wild game markets, as that any merchant, mechanic, or other workman should understand the business, each of his own line, in order to make it to him in any way a paying business. It is only at certain seasons that fur dealers call for furs, and the market does not cry so loud the year round for the venison.

I would recommend for deer hunting a rifle of large calibre, as the hard bone will not then prevent a good shot: it will make a large hole in the hide, and the blood will therefore flow much faster. A breech-loader is the best, for two reasons: it is quickly loaded, and is sure of fire. I have often been asked where I aim at a deer when I shoot one. If it is fair broad side towards me, I aim two or three inches back of the fore leg, one third of the way up the body, being thus sure that the ball, as it enters the body, will go through the heart and lungs.

The hunter should wear, if possible, light colored clothes, as they are less likely to attract attention than dark ones. A cap made of light colored cloth, with a little cape to come down onto the shoulders, is also a very serviceable article of apparel; it preserves the uniform color of the form, and keeps the head dry, in case there is snow upon the trees and bushes. Buckskin moccasins are of twofold service: they are comfortable, and are not noisy.

A dog well trained for deer hunting makes a good companion and a serviceable assistant.

THE BLACK-TAILED DEER.

This deer is found in California, and is sometimes called the California Deer. Those found in the mountains are larger, but shorter than the deer of the eastern part of the United States. Their antlers bear a stronger resemblance to the European Stag than to the Virginian Deer. Its color is reddish brown; a blackish brown streak encircles the chest and shoulders like a collar; the tail is dark brown; the tip of the tail is black, hence the name of this species. It was first noticed by Lewis and Clark, near the Columbia river. It replaces the Virginian Deer west of the Rocky Mountains. It is not as graceful as the latter-named species. The methods used in capturing it are the same as those described for the Virginian Deer.

LONG-TAILED DEER.

There is also in California what is called the Long-tailed Deer. It is smaller and more graceful than the Black-tailed Deer. It resembles the Roebuck.

ELK OR MOOSE.

This species is the largest of the deer family. It stands seven and a half feet high, measured at the shoulders. The horns are palmated, and are very large, frequently weighing sixty or seventy pounds. The weight of its body is from eight hundred to twelve hundred pounds. It is found in Canada and the north-western Lake Superior regions, and is still found in northern Maine. It is very shy. Its sense of smell is very acute. Moose are captured by the still-hunt; torches are used. The hunters cannot go on horseback; as the regions inhabited by them could not be traveled by horseback riders. It is the most easily domesticated of any of the deer.

THE CARIBOU.

The Woodland Caribou, as it is sometimes called, is a variety of the Reindeer. It inhabits the woodland

THE WAPITI, OR "ELK."

and deep snow regions lying between Hudson's Bay and Lake Superior. Its weight varies from two hundred to three hundred pounds.

THE WAPITI OR ELK.

This animal represents in America the common Stag or Red Deer of Europe. It is sometimes called the Carolina Stag, but is popularly known as the Elk. The name Elk is, however, properly applied to the Moose of colder latitudes. The Wapiti or Elk is a native of North America, where it ranges in herds varying in number from ten to several hundreds, and roams from the Atlantic to the Pacific. Its northern range is bounded by the country in which the Moose Deer are found in the greatest numbers. As compared with the Virginian Deer, it is larger, measuring, as it does, more than seven feet from the nose to the root of the tail, and stands about five feet high, measured at the shoulders: in color it is quite similar to the former, and undergoes quite similar changes at different times of the year: the horns are larger, and are even larger than on the English Stag. In their matured condition the horns are of a chestnut brown color: the snags proceed from the anterior surface, the lower-most ones are "looker prongs," so the hunter calls them. What is true of the Virginian Deer is also true of the elk, the number of snags varies with the condition of the animal. An average pair of horns measures four or five feet in width, from tip to tip of each, and weighs from twenty to thirty pounds. The track of the elk is different from that made by the common deer; the latter is quite pointed on its front side, the former is more blunt, being nearly as broad on its front side as on its back. The Caribou makes a track almost like that of the elk, all the others of the family make a pointed track, and in some cases the point is very sharp. The physical appearance, and the habits of the elk are represented by those already given of the Virginian Deer. It is strong, and courageous, and is much more combative in disposition than our common deer. It may be taken by the methods already described for deer hunting.

IV.

THE ANTELOPE.

Antelopes have hollow horns, which vary somewhat in size and shape in the different species. Like the animals previously described, they ruminate: like them, too, many of them have slender bodies, a rapid and graceful gait. In point of physical structure they resemble the deer; but they surpass them in the three points just mentioned. The horns are conical, are bent backwards, are in general not large as compared with those of the deer, are hollow and are permanent and not annually shed and renewed. The hollow, permanent horns form the basis upon which their classification has been made by naturalists. It is stated in the American Cyclopedia that the name of the antelopean family in the Greek, the Hebrew, and the Arabic languages is significant of the brightness and beauty of their eyes. Certain it is, that no brighter or more beautiful eye looks out from the head of any animal than of this one.

The American antelope, or pronghorn, is one of the goat-like antelopes, and is in many respects like the antelope of the Alps, the famous chamois. The horns are erect, the tips only are curved, and are curved inward instead of backward, as is generally the case: there is also a short medial prong. The winter coat of this species of antelope is composed of hollow hairs, which are about two inches long: they are very brittle. The summer coat is smooth and flexible. The pronghorn is found upon the plains of the far west. It is generally captured by rifle shot. It has a great curiosity to know the cause of its fright, when disturbed in any way, and it is generally by taking advantage of this curiosity which causes it to follow up the disturber, that the hunter gets near enough to the animal to shoot it. When apprised of danger, the antelope looks for the cause, and when it sees where it is, advances toward the object by a series of half circles about it, until it has learned what it is.

V.

THE ROCKY MOUNTAIN GOAT.

This goat was formerly classed with the goat family. It is now classed with the antelopes. It is sometimes called the sheep antelope, and the wool-bearing antelope.

It inhabits the highest and most inaccessible peaks of the Rocky mountains, ranging from 40 deg. to 60 deg. north latitude. It is the most abundant on the western slope of the mountains, and the woody country near the coast. In some respects it resembles the nimble, fearless climber of the Alps, the chamois. It wanders over the most precipitous rocks, and springs with great activity from crag to crag. In size, it is like that of our ordinary sheep: in its general appearance it resembles the Merino or Spanish sheep. The wool is fine on many parts of the body, but not as long as that of our domestic sheep. Its chin is bearded. The outer hair, which is long and straight, grows upon the back and upon the top of the head: it is fine and silky, and hangs down like that upon the cashmir goat. Its fleece is a beautiful white. It feeds upon the mosses and grasses which grow upon the mountain sides. Its flesh is dry and hard: it has a musky odor. It has erect, pointed horns: they are small and smooth, and are jet black: the feet are black.

Travelers, who more frequently see the big-horn sheep of the mountain regions, mistake it for the goat. The big-horn lives in the valleys; but the goat rarely descends to the valleys, and never makes its home there.

The Indians make good use of its coat of wool and hair, and of the skin. It is difficult for the hunter to capture it; for many times, when he has succeeded in shooting one, it has been when the animal has stood upon the verge of some precipice, and, on being shot, has fallen over it into a deep and inaccessible ravine below.

VI.

THE AMERICAN BISON.

This animal belongs to the genus Bos, or the Ox family. The family is gregarious in its habits, and is found in every quarter of the globe, tenanting deep glades of the forest, or roaming upon highlands and a "thousand hills." Wild oxen have existed in Europe. The common or domestic oxen, wherever their native home may be, are familiar to every one in their structural formation, and are useful, directly or indirectly, to all classes of people.

Of the three bisons described by naturalists, one is found in Poland, and is known to us as the auroch: another is found in India, but is not well known by any one: the last of the three is the American bison, popularly, but as improperly, known as the Buffalo. So common is this error that but few, in thinking of the buffalo, have depicted upon the mind the humpless, maneless shoulders of the true buffalo of the east, which, in India, has long, smooth horns running backwards and reclining towards the neck, and, in Africa, which carries very massive horns, that are nearly united on the forehead.

The bison is the only species of the ox family indigenous to America.

The original range of the American bison was from shore to shore of what is now the United States, with the exception of a few limited localities: its southern boundary is New Mexico, its northern the Columbia and Saskatchewan rivers. It now ranges in countless numbers and in immense herds over the prairies of the west.

Like the red man, who once roamed with it over all the land, it is doomed to total extinction.

The high arch upon the shoulders, which is covered with very long, heavy, coarse hair, is the peculiar characteristic of this animal. This arch or hump diminishes in height as it recedes: the hinder parts are covered with short fine hair. The general aspect of the bison, as it advances towards one, is that of overpowering strength. It is, however, the most timid and inoffensive creature of its kind, and does not attack men.

The general color is umber-brown, which in winter becomes quite rusty.

It lives upon grass and other herbage. In the winter it might often suffer from extreme hunger, but that the nose is peculiarly adapted to shoveling away the snow which covers the grass. It is a good swimmer, and in the summer resorts to the water for bathing.

Like the deer, it delights in visiting salt-licks. The flesh is considered by many very palatable.

The fur, the hair, the hide, and the tallow are all important articles of commerce. The Indians depend largely upon them for food, clothing, and shelter. The weight of the bison is sometimes as great as two thousand pounds. They migrate in herds of hundreds and of thousands, and some affirm that they have seen as many as twenty thousand in a herd. The herd follows its leaders with such determination that it is almost impossible to turn or arrest its course.

Hunters take advantage of this, and, when it is possible, drive these creatures over some precipice. In this way hundreds are killed at once. Indians take them by this method. They likewise use the bow and arrow. White men take them by firearms, shooting while on the chase.

PRAIRIE-DOGS.

VII.

THE PRAIRIE-DOG.

The Wish-ton-wish, or prairie-dog, as it is popularly known, is allied to the marmots, resembling them in its physical structure, and in many of its traits. The burrows which it makes are like those of the common woodchuck. Its name is due to the short yelping sound which it utters, and which is so like the bark of a young puppy. It is found in great numbers along the course of the Missouri and Arkansas rivers, and also near the river Platte. In the vicinity of Potter, a station on the Union Pacific Railroad, four hundred and thirty-three miles west of Omaha, the observing traveler will begin to make acquaintance with the prairie-dogs or cayeutes, as the Indians call them; and at about three miles from the station is situated the great prairie-dog city, occupying several hundred acres, honey-combed by a perfect labyrinth of subterranean burrows.

Wood says of them:—" The scene presented by one of these 'dog towns' or 'villages,' as the assemblages of burrows are called, is most curious, and well repays the trouble of approaching without alarming the cautious little animals. Fortunately for the traveler, the prairie dog is as inquisitive as it is wary, and the indulgence of its curiosity often costs the little creature its life. Perched on the hillocks, the prairie-dog is able to survey a wide extent of horizon, and as soon as it sees an intruder, it gives a sharp yelp of alarm, and dives into its burrow, its little feet knocking together with a ludicrous flourish as it disappears. In every direction a similar scene is enacted. Warned by the well-known cry, all the prairie-dogs within reach repeat the call, and leap into their burrows. Their curiosity, however, is irrepressible, and scarcely have their feet vanished from sight, when their heads are seen cautiously protruded from the burrow, and their inquisitive brown eyes sparkle as they examine the cause of the disturbance."

It is difficult to get these animals even after they are shot. Tradition says of them that when one has been shot at the entrace of the burrow, those inside draw it out of sight and reach. I have seen Indians catch them. They shoot them with the arrow, and in such a way as to have the arrow point reach through the animal and stick in the ground, thus pinning them fast and preventing escape on their own part, and likewise preventing escape by the help of another dog.

VIII.

THE WOOD-CHUCK OR MARMOT.

The Marmots are found on both continents; but in this country they are commonly called Wood-chucks. They are found from Hudson's Bay to South Carolina, and west, to the Rocky Mountains. They are about the size of an ordinary rabbit, and resemble it in color, being blackish or grizzled above, and chestnut red below. They are clumsy looking fellows, slow and awkward in their movements. The head is broad and flat; the neck is very short; the legs are short and thick; the feet are large; the tail is bushy; the hair is quite soft; the whiskers are long and stout; the eyes are small and the ears are short; they have rudimentary cheek pouches; the stomach is simple. They are expert excavators, and dig large and complicated burrows in the fields, on the sides of hills, or under rocks in the woods. The burrows slant upwards, to prevent the entrance of water into them: there is usually more than one entrance to a burrow. They feed mainly upon vegetables, and are particularly destructive to red clover crops. They are very clean animals. Though naturally timid, when they find themselves unable to escape, they will fight desperately, and, with a dog of equal size, with great success. They are sometimes called ground hogs. They are gregarious. They frequently make their incursions upon clover fields, or other forbidden grounds, at midday, posting sentinels to warn them of danger. This the sentinels do by a shrill whistle:

they are very vigilant, and as the sense of hearing is acute, they can easily manage an escape.

They are with little trouble caught in steel traps set at the entrance of the burrows. The traps must be concealed, but need no baiting. The skin is valuable for whip-lashes.

THE PINE MARTEN.

IX.

WEASELS.

The members of this family have very long, slim bodies. Their legs and feet are very short. Their five toes, which have sharp claws, are short and round. They are great climbers. The structure of the teeth is such as to fairly rank them among the carnivora. They eat small quadrupeds and are specially fond of sucking the blood of their victims. They have a gliding, almost serpent-like motion, and this, together with the slenderness of their bodies, often gives them a chance to steal unawares upon animals, which are greatly their superiors in size. They generally attack their prey upon the neck or head; and breaking the skull, they drink the blood, and then frequently leave it otherwise untouched. When they are frightened, or in any way disturbed, they emit a very disagreeable odor. Were it not for the value of their fur, we would scarcely care to be intimately acquainted with the offensive and troublesome creatures. They are nocturnal in their habits; a few members of the family will sometimes venture out into the light of day in search of food.

"Fine plumage does not make a fine bird," neither does the sable nor the mink skin make the weasel family a good natured, well mannered class of beings.

PINE MARTEN.

The pine marten, first on the list in the weasel family, is found in those regions which abound in pine trees.

It is frequently called the American Sable, representing, as it does, the sables of the old country. It inhabits both the old and new world. The sables inhabit Russia, Japan, and Northern Asia. Their skins are the most valuable of any of the weasel family; the Russian Sable being the most valuable of the species. The American Sable, or Pine Marten, ranks next in value; then the Beech Marten follows in the order, and last and least of all, the American Fisher.

The Pine Marten of Hudson's Bay locality is often called the Hudson's Bay Sable. It is by nature very cautious and timid, and does not often approach settlements. Its native locality is the high ground of thick pine woods. Dr. Richardson observes that "in America, particular races of martens, distinguished by the fineness and dark color of their fur, appear to inhabit certain rocky districts. The rocky, mountainous, but wooded region on the north side of Lake Superior, has long been noted for its black and valuable marten skins." It varies the most in color, in the same region, of any animal; some are light yellow; some are dark brown; and others are black. It is distinguished from the Beech or Stone Marten of Europe by the coloring of its throat. The former has a yellow throat, the latter a white throat, and is sometimes called the White-throated Marten. It feeds on small quadrupeds, and is very fond of nuts and honey. The Pine Marten is about twenty-eight inches long, including the tail, which is ten inches in length. It is a great traveler, but scarcely ever makes a straight track: this is either quite crooked, or is curcuitous. Because of its irregular course, sportsmen and inexperienced hunters are apt to think the woods are full of martens, when, in fact, there may be but very few of them. It is very essential for the young hunter to learn to distinguish the tracks of the animals that he wishes to capture. The Mink, Marten, Otter, and Fisher, travel about alike; they make the same motions, and the shape of the tracks is very similar. The marten moves along after the fashion of a rabbit, only it leaves but two tracks. These tracks are about two inches apart, and the one is not exactly in front of the other. It springs eighteen or twenty inches at a time. Its track is a little larger than that of the mink, for its foot, being covered with fur, spreads more. They are otherwise alike.

In sections where there are but few martens, they can be followed up, by their tracks, to hollow trees. When overtaken, they will hardly ever go into the tree which they have climbed, but will go from it into the hollow tree. This is cunning; but the hunter can match it by his knowledge of the appearance of the timber of a tree that is hollow.

The marten is a great climber, and spends much of its time on the trunks and amid the branches of trees. It is very sprightly in its motions; its muscular powers, as is also true of other weasels, are wonderfully perfected, and it is more than a match for any squirrel. It is also a silent traveler, and by stealth comes unnoticed upon its prey. It delights in robbing nests, taking from them eggs, or the young or old birds. It likes, after having removed the occupants of a nest, to appropriate it to itself, thus securing comfortable quarters, and a good chance to watch the birds in nests close by. Sometimes it drives a squirrel from its burrow, and inhabits it. In the winter it prefers to live in the clefts of rocks, or in hollow trees and logs. It is very fierce when attacked, or even disturbed. When a man approaches a tree where one is secreted, though he may be ignorant of its presence, it will come out to fight him. It will even come from the farther end of a log or cavity in the tree. It is safe to say that it would come forty feet to attack one. On one occasion I found a marten in a hollow tree, and thought I would stop up the entrance and afterward go and kill it. Hunters very often catch them in this way; but this little fellow had no idea of being caught in such a trap: he came out to fight me. I then held my gloves in the little doorway, while I fixed my gun on a log near by; as I drew them out, the marten came too, and looked me in the face while I shot him.

It can be taken in dead falls, set one quarter of a mile apart, and can also be taken by the steel trap. It is very easily caught. When steel traps are used, the hunter makes what he calls marten trails; a track that it might take two or three days to follow round. The traps are set on this line. A coop should be made to put the trap in; the bait should be put in the coop, beyond the trap. Hollow logs and trees are excellent places for trapping. Any kind of meat, especially if bloody, is good for bait; the more bloody the meat, the better the marten likes it. The musk of muskrat is good bait; the heads of birds and fishes is very good. The methods used for trapping mink are further discriptive of marten catching.

It is a very handsome animal, bright and sprightly, and would make a very choice pet if it could be tamed; but it cannot.

THE MINK.

The northern parts of Europe, Asia, and Africa are the homes of this handsome little creature. Its fur, especially in the United States, is very valuable. It is more carnivorous than the marten, its dentition being

THE MINK.

somewhat different. It is smaller, more slender, and more uniform in color than the marten. The northern species bears the darkest, finest fur. Its color is a dark brown, with a strip along the back of a deeper shade; it has a patch of white on the under jaw. It is probable that the climate affects the color of the mink in different localities. In size mink vary from thirteen to eighteen inches from the nose to the base of the tail. The tail measures from eight to ten inches in length; its color is nearly black: it is also quite bushy. Mink are larger in their north-western homes than any where else. The feet are slightly webbed, thus adapting the animal to the water, which it frequents. It is an excellent swimmer and diver, but it does not stay long in the water. Hunters look for mink along the banks of streams. It is a good runner, but does not climb like the marten. It is a great rambler, except in the breeding season, which commences about the last of May. The mother hides her four, five or six young ones until they are half grown; for the males of the family, inhabiting the same localities, like to destroy them.

Its food consists of fish, frogs, and birds; it is specially fond of speckled trout.

It is a good plan to catch mink before it is in the best condition, as it is much easier caught, and will be more uniform in color. In the Lake Superior regions it is in its best condition about the 28th of October, in southern Wisconsin not before the latter part of November. Two can be caught easier before they are prime, than one afterwards. They may be caught in box traps, and kept and cared for until such time as the hunter may choose to kill them. They are usually caught by the steel trap, set either on the land or in the water. Sometimes a hole is dug in the snow, three sides of which are barricaded; the trap is set at the entrance of the hole; the bait is put in it, beyond the trap. This little barricade, or coop, should be covered over with evergreen boughs, to keep out snow and also other stuff that in falling might spring the trap. The trap itself should be concealed by a covering of leaves; rotten vegetation makes a very good covering. The flesh of the muskrat, and its musk will attract the mink for a long distance; fish oil is good bait. This oil is obtained by the decomposition of the fish, which is accomplished by exposure to the sunshine. Trappers often set the trap in the entrance to one of their burrows, or, if such a hole cannot be found, they make one by the side of a root, or stump, or on a bank. It must be nicely set; as the animal is suspicious of what is new and strange. The trap may be set in water, covered about two inches by it; the bait is put on a stick about eight inches beyond the trap, to oblige the animal to walk over it. The mink likes to step out upon fallen timber which lies over, or one end of which lies in the water; because from such a point it can watch and dive for its finny prey. The trap may be set in a hole made in the end of such a log, or put upon it and covered with the moss that so frequently gathers upon such places; the bait is then put beyond the trap. Though the animal may not be hungry, it will go to smell of the meat, and thus decide its fate.

THE FISHER.

This member of the weasel family, which is sometimes called Pekan, inhabits many sections of the United States, from North Carolina, on the south, to the Great Slave Lake on the north, and across the continent, from shore to shore. It is common in New York, Pennsylvania,

THE FISHER.

and the Lake Superior mineral regions. It lives mainly in damp localities, and in humid forests bordering water.

It is a carnivorous digitigrade, measuring about two feet in length, except the tail, which measures fifteen inches. It weighs from eight to twelve pounds. In color it is blackish, with a greyish tinge about the head and shoulders. Some are brown, and some have a white spot on the throat. In general its appearance is fox-like, having a long head and a pointed muzzle. It is the most ferocious member of its family. It preys upon birds and small quadrupeds, and, like the mink, it is often seen on the end of a log lying over the water, ready to catch the fish as they swim along. We do not know that it receives its name from any special liking to fish food: the name is more likely to have been given it because it delights in stealing fish bait set for martens.

It escapes their traps. It must be severely wounded to be caught. The hunter, therefore, sets stouter traps, and makes heavier dead-falls for it than he does for the marten. It is easily caught in traps. The best and the easiest method is to set fisher traps and marten traps in lines near each other, taking the lower ground for the former. These two animals are in many respects very similar. They inhabit the same localities; but the fisher travels on lower ground, and in straight lines. Traps for the fisher are set in hollow logs, or in barricaded places, such as have already been described, but the barricade and the entrance to it must be larger than for the marten or the mink. The trap must be fastened, by a wedge or stake, into a tree. It is an excellent plan to draw a trail of the musk of the muskrat, mixed with fish oil, to attract the attention of the animal. This scent may be put in a deer's skin bag, which has been pierced in a number of places with an awl. The fur is generally sent to England.

THE BADGER.

By many good authorities this animal is placed in the weasel family; although in some respects it resembles the bears. Its dentition is not like that of the ursine family. It is an omnivorous plantigrade. Like others of the mustelidæ, or weasels, it feeds upon small quadrupeds, and it also eats roots and fruits; the latter is said to be its choice of food-stuffs. It plunders the nests of the wild bee, eating with evident delight the store of honey, fearless of the sting of the bees. The structure of the skin is such as to render it impervious to the attacks of the little sharp-shooters, and it will seem as much at ease in the midst of the besieged army, as does the bee-tamer surrounded by his stupefied throng.

Its feet are five toed, and are well adapted to burrowing: the powerful claws are deep set in the flesh: the claws of the fore feet are long and curved. The muscles of the legs are wonderfully developed. With the fore feet it digs rapidly, and can dig very deep; with the hind feet it can as effectually fling the dirt. It frequents deep woods, where it digs its burrows, and in which it sleeps during the day. The burrow is curiously and conveniently constructed. It has but one entrance, but it has several apartments, the innermost one of which is circular. The animal lines these apartments with grass, or other soft material. It spends its days alone, a solitary creature. It is cleanly, timid, and inoffensive, but, if attacked, becomes very fierce in self-defense. If attacked when in its burrow, it will fling dirt in the face of its invader, and frequently with such vigor as to make very sure its own defense.

THE BADGER.

It is slow and clumsy in its motions, and in this respect resembles the bears more than the weasels. Its skin is very tough: it is valuable in commerce: it makes excellent pistol-holsters. The hair is used in various ways. In color it is a greyish brown, a curious intermixture of brown, white, and black. In the summer the color is more of a yellowish brown. The winter fur is thick and handsome; it measures about three inches in length. The average length of the badger is two feet and a half; its height at the shoulders is eleven inches. The European species is the most important in the fur trade, furnishing a large majority of the skins which are annually sold.

The badger can be caught in the steel trap, set at the entrance to its burrow. It should be carefully concealed, else the animal will cunningly avoid the suspicious looking thing. It may be baited with fresh fish, or with salt cod-fish. If the latter is used, it should be roasted, to give it a strong smell. Wood, in his Mammalia, gives an instance to prove that the badger is not as stupid as it is generally supposed to be. He says, "One of these animals has been known to set at defiance all the traps that were intended for its capture, and to devour the baits without suffering for its temerity. On one occasion the animal was watched out of the burrow, and a number of traps set round the orifice, so that its capture seemed tolerably certain. But when the badger returned to its domicile, it set at naught all the devices of the enemy, and by dint of jumping over some of the traps, and rolling over others, gained its home in safety." I have known them to elude the jaws of the trap by covering them over with dirt, putting on so much that it could not spring when stepped upon.

THE OTTER.

The otter is the aquatic representative of the weasel family. Many of the weasels resort to the water; but none are so thoroughly at home in it, and none so poorly adapted to spend its time continuously upon the land. It bears some marked resemblance to the seals. There are several species, which are distributed quite extensively over the globe. The two species more specially interesting to the American trapper are the Canada, or American Otter, and the California Otter. The range of the former is bounded on the north by the Arctic ocean, on the east by the Atlantic, south by the Gulf of Mexico, and on the west by the Pacific ocean.

At a little distance it presents the appearance of a very large mink. The American Otter measures four and a half feet from the nose to the tip of the tail. It weighs from twenty to twenty-five pounds. The shape of the head differs considerably from the other members of its family, being very broad and flat above, the outline of the muzzle being round. The eyes are small, and set far forward, and are provided with a nicitating membrane; the lips are large and fleshy, and are provided with strong whiskers; the tongue is rough; the ears are short and round; the teeth are very strong, pointed, and sharp; the fur is smooth; the under fur serves as a protection from the extremity of the weather, and together with the long, glossy hairs constituting the outer fur, indicates the aquatic habits of the animal. The color of the fur is a bright, rich brown, varying somewhat with the locality, and the light in which it is viewed; the color on the upper side is a dark, glossy brown, and on the under side is much lighter. The sides of the head and throat are covered with a dusky white fur. The tips of the inner fur are brown, while the base is grey.

When on the land it is plantigrade, and is a very awkward, clumsy traveler. The toes are webbed and spreading; the legs are strong, and are so constructed as to enable the animal to move them in almost every direction. The tail, which serves as a rudder, is long, and depressed at the tip, stout and muscular at the base. The otter is a very graceful and rapid swimmer, swimming at every depth with perfect ease. It glides along so quietly that when in deep water it scarcely moves the surface by a ripple. When in deep water its course may be determined by the bubbles of air, which it comes to the surface to exchange for fresh air. It is persistent in pursuing its prey; its slim body gives it great advantage in following through every turn and winding way. It swims in clear and rapid streams, the homes of its favorite prey, the speckled trout.

The otter selects the finest fish, sometimes bringing to the shore, or upon some rock which rises above the surface, or to some half sunken log several fish at a time. In eating them it begins by crushing the head between its teeth; it then eats the flesh of the body of the fish. Some affirm that the otter seldom eats any part but the head; others that it never eats the head nor tail of the fish. My own observation leads me to believe that the degree of hunger, and the kind of fish caught are the influences controlling the amount and parts of fish devoured by the otter. I have never seen any part of the speckled trout left, nor any part of any fish, except the heads of bass; but I have often seen whole fish left, and I know that the otter often brings up more fish than it even attempts to eat. It is more greedy than fastidious. Like the drunkard, whose appetite required more than his constitution could bear, the otter's greed demands more than enough to satisfy its hunger. Two years ago, while trapping mink in the Lake Superior regions, I one day saw more than a bushel of bass heads lying on the bank, which otter had left there, and which the mink were carrying away.

It is a great rambler. It will sometimes be gone for days from its home. It finds along the banks of streams knolls, or high places, which are inclined towards the water. At all of these places it stops and plays; it rolls and tumbles about upon them. These high banks are called "otter slides." They generally project into the streams, and are near deep water, where fish are abundant. The animal slides from these to procure its food, and it is probable that from such points the young otter learns to slide so gracefully and noislessly into the water. Certain it is, that it does at some time acquire both a noiseless and graceful descent into the water. When it comes onto a slide from the water, it does not come directly upon the slide, but approaches from one side of it where the water is shallow and the ground low. It seldom travels any great distance on land. The slides are almost always located on the bends of streams, and it will sometimes go from one slide to the next one by land. The paths which the animals make hunters call otter portages. The otter is so sly that if a person goes onto one of its slides, or portages, it will leave it, and will frequently abandon it entirely. They travel in groups of three six, or seven, but generally in groups of three, of four, or of five. They will go from one to two miles per day, stopping at their slides for play spells, unless these places have been disturbed. They will frequently portage from the head of one stream to the head of another. I have known them to do so when the streams were two or three miles apart. The otter's track is very much like that made by the fisher; but it may be distinguished from the fisher's track by this, that it frequently slides along for a ways, sometimes for six or eight feet. The otter also leaves what is called a "seal," or an impression of the sole of the foot, by which its track may be distinguished from that of any other animal.

It burrows in banks, or hides in natural crevices. It will sometimes steal muskrat burrows.

As a rule, it is unadvisable to shoot fur-bearing animals; because it injures the fur. The otter and all other aquatic animals are almost invariably lost if killed by shot; for they will then sink.

The otter secretes a strongly fetid substance, and may therefore be readily tracked by other animals. Otter hounds are frequently used for catching them.

The steel trap is the best instrument of capture for this animal. A great many trappers set their traps on the land, counting the slides the best places for them. I do not recommend this course. For the practical trapper it is a waste of every expenditure of effort. It is true, there may be cases when it is better to set the trap on the land, instead of in the water; but it is not often the case.

When I first went into the Lake Superior regions for trapping, some of the men, whom I found there, always set their otter traps on the land; but I at one time found that I had caught fourteen by water trapping while one of the men had been waiting for a single catch. When a trap has been set on land, the scent of man will not leave the vicinity for days, and so long as the otter discovers this scent, it will not approach close to it, and if it has been several times towards the spot, is quite apt to abandon the place entirely. Besides, if the animal were not so afraid of man's tracks, it were still an unadvisable course to pursue; for it is frequently gone from its home for a week at a time. The trapper, knowing this, must not be surprised if his trap is not sprung for several days, and should remember that it is at best a slow process. No bait is needed for the trap, and no scent should be left. There is no scent left if the trap is put under water, and many times the animal may be caught within an hour after the trap is set. The trap should be set near a slide, on one of the paths by which it approaches a slide. It should be covered by four inches of water, enough to keep off the muskrat. It should also be set about four inches to one side of the center of the path; because the legs of the otter spread out so far from the sides of the body that it makes a wide path, and the tracks of the feet are on the edges of the path. If the trap is set on the center of the path, the body of the animal passes upon it without injury, but if to one side, as stated above, the trap is sprung and the feet caught. The trap should be carefully covered, to prevent suspicion.

X.

THE SKUNK.

Five species of the skunk are found in North America; the White-backed, the Long-tailed, the California, the Little Striped, and the common skunk of the United States. The skunk is found east of the Missouri plains, and from Hudson's Bay to Texas. This animal bears many of the typical characteristics of the mustelidæ or weasel family, having an elongated body, pointed, naked nose, fossorial feet. The claws of the fore feet are long; the soles are naked; the tail is long and bushy; there are five closely united toes. It is essentially plantigrade, and walks with the back much arched, and with the tail dragging, except when surprised, when it erects the tail. The head is small; the eyes are small but piercing; the ears are short and rounded. It measures from sixteen to twenty inches in length, the tail being thirteen or fourteen inches additional. The prevailing color is black and white, or black with narrow white lines on the forehead, a broad triangular patch on the nape, continuous with a narrow line on each side of the back; the tail and tuft is white. Of the teeth there are six incisors, and two canine teeth in each jaw, eight molars in the upper, and ten in the lower jaw.

It is nocturnal in its habits, and during the summer months it feeds upon beetles, grasshoppers, crickets, and other small animals, upon eggs, upon green corn and other vegetable food. It preys upon hens and chickens, but not so frequently as does the mink or the weasel. It burrows from December to the middle of February: it carries with it no winter stores. It is fat when it retires to its burrow, and remains dull and inactive while there, but cannot properly be said to hybernate. Its flesh is white, fat, and when properly skinned is not tainted. The Indians eat it. Though slow in its movements, and weak and timid, it is by no means defenseless. It furnishes the staple fur in Poland, and is therefore valued by the Polanders; but it is the most thoroughly detested animal in the United States, and as the wit has well said, it is literally in " worse odor " than any other animal known. Its fetid secretion is its sure defense against any molestation, and there are few who would thank us to tell them what everyone already knows of this disgusting, all-permeating, ever-penetrating substance. The skunk is, nevertheless, a very cleanly animal; as it never allows its own fur to become soiled by its secretion, nor does it emit it upon its fellow skunks.

It may be taken in traps. These traps, which should be loosely covered with grass or other soft stuff, may be baited with pieces of meat scattered about it. The traps should be set on one of its paths where it searches for food, or enters a burrow. It may be taken by the snare and spring pole. It seldom emits its fetid secretion until the hunter attempts to kill it. If the animal is immersed in water no scent will be emitted.

XI.

THE WOLVERINE.

The Wolverine or Glutton, as it is frequently called, is a connecting link between the weasel and the bear families. It resembles the former in its dentition, the latter in its plantigrade character. In many of its habits it resembles the martens. The glutton and the wolverine were once supposed to be distinct species, the glutton inhabiting portions of the Old World, and the wolverine of the New World. They are now regarded as belonging to the same species; the differences between them are doubtless the result of climatic influences. It is a native of the high northern latitudes of Europe and Asia, and of the cold regions of America. Its American home extends as far south as the mountains of Massachusetts. It is most abundant in the Rocky Mountains near the Arctic circle, and in the west is found as far south as Great Salt Lake; it has been occasionally found in northern New York.

It has thirty eight teeth, which are peculiarly predaceous in their character. It is weasel-like, as its teeth would indicate, in its choice of food stuffs. It feeds upon small quadrupeds; but being larger than some mustelidæ, and more voracious, it seeks larger prey than weasels do. It is a persistent foe to the beaver during the summer months. It feeds upon the carcases of animals which it has not killed. Early writers of the Eastern Continent in describing this animal put its strength, voracity, and cunning in the superlative degree. The gluttony which they attribute to it gave it the name glutton. It is now believed by those whose practical knowledge of this animal should be accepted as important, that this characteristic was exaggerated. It is not probable that by force of a natural apetite, or by dint of agility and cunning, it pounces upon deer and other large game, or that it plunges into water in search of game ; but it may, and probably does steal upon the deer asleep, or attack fawns or feeble animals, and destroy its victim as the weasel does when it attacks a hare, by striking at the blood vessels of the throat. Upon this point all writers and hunters agree, that the wolverine is very annoying, because it will rob traps. It has been known to rifle the marten hunter's path round a line of traps extending upwards of fifty miles. In most cases it only disturbs the traps by devouring the bait. It does not like the marten, but often tears it to pieces and conceals it near by: the keen scented fox, greedy for such game, follows up the track. It is therefore well to set fox traps near the marten traps. The wolverine delights in plundering the hunter's "caches" of provisions, eating the meat and scattering the vegetables. The Indians, because of its destructiveness, call it the devil.

It is active all winter, nocturnal in its habits, and spends the day in holes and caves. It fights resolutely, and is more than a match for a very large dog.

The general contour of the body is like that of the bear. It is partially plantigrade; its track resembles that of a bear's cub; its gait is slow, but what it might lose by this defect it gains by its persistent, steady pace. It measures two and one half feet, exclusive of the tail, which is about eight inches long; the tail is bushy. It is fifteen inches high at the shoulders. The width of the hind foot is four inches; there are five deeply divided toes; the claws are long, sharp, and curved; these are sometimes worn by the Indians as ornaments. The eyes are small, of a dark brown color. The general color of the fur is a very dark brown on the back, changing into black on the limbs. There are bands of lighter brown passing from the neck to the flanks and meeting at the tail, and also one crossing the forehead from ear to ear. The long fur is fine and glossy, and is quite valuable. It is used for sleigh robes. The weight of the animal is from twenty to thirty pounds.

Hunters generally set poison for it. It can be caught in the steel trap, and in dead-falls; the latter should be very heavy. In wolverine countries every tenth trap on a marten line should be set for wolverines.

XII.

THE BEAR FAMILY.

This family, as to point of numbers, is not numerous; but the individual members of it are notably large. They differ from the typical carnivora in several respects; they have massive limbs and a heavy gait, and depend less upon animal than upon vegetable food. A full grown bear of the majority of the species generally weighs several hundred pounds. The extremes in weight are twelve hundred, and less than one hundred pounds: the former designates the weight of the Polar Bear, the latter that of the Bornean Bear.

Bears are not very aggressive, and with a few rare exceptions, are singularly harmless. Ignorance of this fact is not bliss to the many who would not for their lives camp out a night in or near woods where bears were supposed to be. They rarely attack a man if undisturbed by him. Many of them are timid, and may be quite easily frightened off when one has come unexpectedly upon a man's track. A partner of mine once frightened an infuriated bear, from which he had just taken her two cubs, by making a lunge at her with an open umbrella. She ran off in hot haste, and was not seen again. Children are taught by some one a most pernicious fear of bears, thousands of whom never see one uncaged, and yet in after years they remember very distinctly their childish fears of them. They recall with a smile the "Let's play the bears are going to eat us up," and with indignation that when night came on, or they were shut up in some dark closet for being "so naughty," they scarcely dared to move a muscle, or the eyes from the opposite wall where the veritable bear they had heard so much about, sat like some demon watching them, and just ready to devour them. Naughty bears do not eat naughty children now; nor do the wild bears of the woods eat the big babes of the wood who know how to use traps and rifles. The Scandinavian aphorism, "A bear has the strength of ten men and the sense of twelve," need not frighten the woodsman, but should teach him that he needs to learn how to manage this creature. An infuriated bear is a formidable antagonist.

Bears belong to the order omnivora. Destitute of the powers which distinguish the carnivora, they still exhibit the same natural adaptation of endowments and wants. With a few exceptions they content themselves with a vegetable diet, living upon fruits, nuts, and succulent vegetables; they are almost proverbially fond of honey. They are very persistent, and cunning too, in their efforts to secure bees' nests. They tear the bark from old logs to get snails and insects therein secreted; they tear up old logs to get the insects and worms lying underneath; they gnaw into old stubs, dig holes in the ground in search of wasps' and hornets' nests. They are very fond of fish, and have still greater relish for skunk cabbage. Scarcely any other animal will eat this cabbage. They eat wild or pigeon cherries with the greatest delight. All the bears eat animal food; but they do not slay their victims by attacking them in some vital part, as do those more strictly carnivorous animals; but they hug or tear them to death. One of the endowments which adapts them so well to their food-hunting is their plantigrade structure. Their massive paws easily crush small animals upon which they feed; the claws, which are more tractile, make, like the canine teeth of the lion, the best instruments for tearing the flesh of animals, and make good excavators of roots, of which they are fond. Because of the large surface covered by their feet when placed upon the ground, they are capable of erecting themselves on their hind feet, and of supporting themselves with the greatest ease in an erect position. By this means they are enabled to gather nuts and berries. They are excellent climbers. The feet are five toed. The vegetarian character of these animals is indicated in part by the dentition; the molars have tuberculous crowns. They have twelve incisors, six on the upper jaw, and six on the lower.

Bears are nocturnal in their habits. They hybernate. During the fall their favorite food stuffs are the most abundant; the bears partaking greedily of them become very fat; the honey, which they delight most of all to eat, like other sacharine substances, is a great fat producer, and containing so much carbon as it does is a great heat producer. As people who are in crowded assemblies, in poorly ventilated rooms, become stupid in proportion to the amount of carbon which they inhale, so do bears fit themselves most thoroughly for their lethargic state by the carbonized food which they have eaten. They are careful to make warm beds for themselves of dried leaves or small twigs; or they find a shelter under some rocks, or by the roots of some tree. The hunter says of the bear, that he goes into his burrow "full fat," and comes out "full fat." This is the case. He comes out as fat as when he begun his hybernation; but he becomes rapidly lean when he begins to travel in the

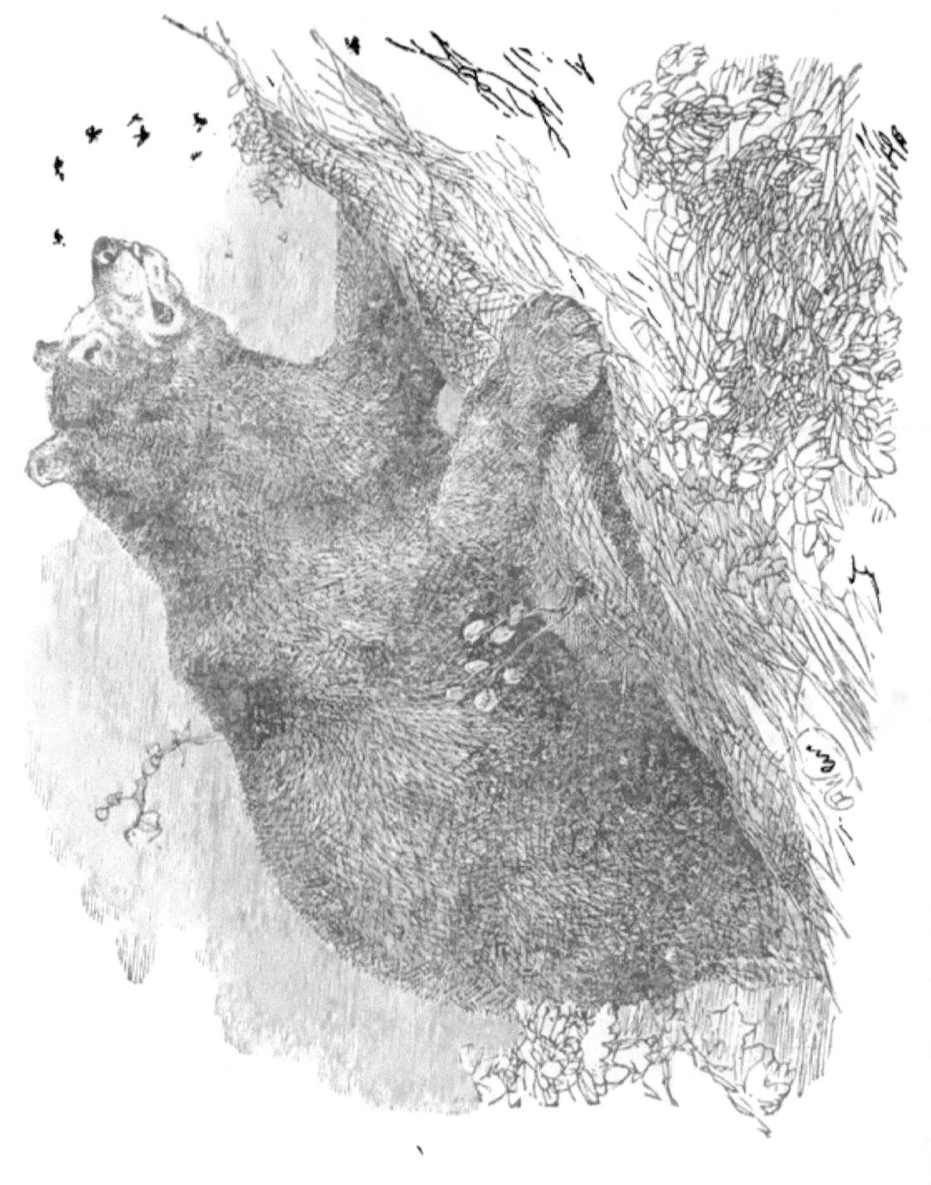

spring. A curious phenomena takes place in the digestive organs when the bear lies down to his long sleep, which prevents any injurious effects from coming upon the stomach in its long state of inactivity. In the summer bears live in some burnt district, in thick timber, among fallen trees, or blackberry brush. On the frontiers they frequently depredate the pastures and barn-yards. They will climb any kind of tree large enough to support them. "Bear-baiting" with mastiffs was formerly a royal amusement in England.

They generally bear two young ones in February or March. The mother hides her cubs until they are large enough to follow her: they remain with her until the following spring.

Bears differ from each other more than the members of almost all other families; because their range of country is greater. They are found in the three zones; those of the torrid and those of the frigid zones are as perfectly adapted to the climatic extremes as are those of the temperate zones to their intermediate condition. They are found on both continents. The American Black Bear and the Grizzley Bear are the most conspicuous of the several species found here.

BLACK BEAR.

The Musquaw or Black Bear is the common bear of America. "It inhabits," says Dr. Richardson, "every wooded district of the American continent from the Atlantic to the Pacific, and from Carolina to the shores of the Arctic Sea." It receives its common name from the color of its handsome glossy fur. This fur and the skin and the fat of the animal are in so great request that its numbers have been greatly reduced. Like other predaceous animals it has been hunted as a nuisance, and driven away from thickly settled localities. The vast forests which have never yet echoed to the sound of the pioneer's axe or rifle, sparsely settled regions, and mountain ranges are the places of its abode. They will continue to be; but as "westward the star of empire takes its way," and as men are constantly leaving old homes for other and newer ones, so will the Black Bear move on before the civilizers of new and unknown lands.

It is closely allied in its habits to the common Brown Bear of Europe, but is less fierce and sanguinary. Its weight is from two hundred to four hundred pounds. It rarely attacks a man unless provoked, save in the case of a female with cubs, the retreat of which she is solicitous to recover. The females with young secrete themselves so carefully that hunters scarcely ever discover a pregnant bear. The bear leaves no track by which the hunter can tell exactly where it may be found; but there are various signs in the woods by which he can tell about where it is, such as rolling logs, gnawed trees, and in the fruit season brush piles. Hunters can tell if a bear is in the hollow of a tree by the marks it leaves upon the bark of the trunk. In ascending it leaves only the puncture of the claws; in descending it makes long scratches. Bears climb trees to "lap," as the hunter says, by which he means that they draw in boughs to get the fruit.

Bears are often hunted with dogs. This practice is ill-advised, because often fraught with so much danger. When a dog is trained to bear hunting, it enjoys it. The hunter should train his dog to attack the bear from the rear, and to catch it by the hind leg; else, if the dog rushes upon it from the front, the bear will raise itself upon its haunches, and its best implements of warfare, the fore feet, are free and ready for combat. A bear, when long worried by a dog, will climb a tree if possible. Sometimes it will come down from the tree, run a ways and "tree" again.

Horses are almost useless in the bear chase, as it is difficult to train them for it, and unless they are well trained their fear lays them at the feet of the bear, and leaves the rider in an almost defenseless condition. When the bear is wounded or terribly frightened, it fights with desperation, and nothing but rifle shot checks its fury. By whatever means a bear is captured, it is imperitive that the hunter understand himself before he attempts to carry off his prize. Many times when the animal is very near its death, hunters have approached their game with confidence, but have found when it was too late to save themselves that the bear was not dead, only dying, and with the last agonies of death it has turned upon the hunters and killed them. The last struggles are the most desperate ones. For the sportsman such hunts may do; but for the practical woodsman there are better methods.

The easiest way to catch the bear is by a trap. Steel traps are used effectually, and as described by Newhouse; "In trapping for bears a place should be selected where three sides of an enclosure can be secured against the entrance of the animal, and one side left open. The experienced hunter usually selects a spot where one log has fallen across another, making a pen in a V shape. The bait is placed at the angle, and the trap at the entrance in such a situation that the bear has to pass over it to get at the bait. The trap should be covered with moss or leaves. Some think it best to put a small stick under the pan, strong enough to prevent the smaller animals, such as the racoon and skunk, from springing the trap, but not so stiff as to support the heavy foot of the bear. The chain of the trap should be fastened to a clog." I would put a small stick over the first spring across the trap. The bear will step over this stick; because it will step into the lowest place in the track. The trap should be put on one side of the center of the entrance; the dirt should be dug away so that the pan of the trap shall be the lowest spot in the path to the bait. It is best to leave several pieces of bait; if the trap is not touched the first time it will surely be at another time. The clog for the trap should weigh thirty pounds. The chain should be short; because the bear in attempting to extricate itself from the trap, swings its foot against a tree or log, in order to break the trap; if the chain is short the force of the swing of the foot is broken. A bear in advancing upon an antagonist rears itself upon its hind feet, and walks with the body erect, the fore paws ready for use as soon as within reach of the foe. It will sometimes walk backwards with the body erected. On one occasion a bear got away from me, and carried off my trap, walking backwards as she went from me. The clog was twelve feet long, five inches thick at one end and seven at the other. It was a dry maple pole, and about eighteen inches of it was put through the ring of the trap chain. The bear was going through a windfall; the timber there had been burned over, and the wind had blown down what trees were left standing. These trees and logs were lying crosswise in every direction. The clog would catch into the fallen timber as she went along; to avoid this she lifted her fore foot and held the trap in her mouth. She walked backwards doubtless to preserve her equilibrium; for had she attempted to walk forwards the weight would have pulled

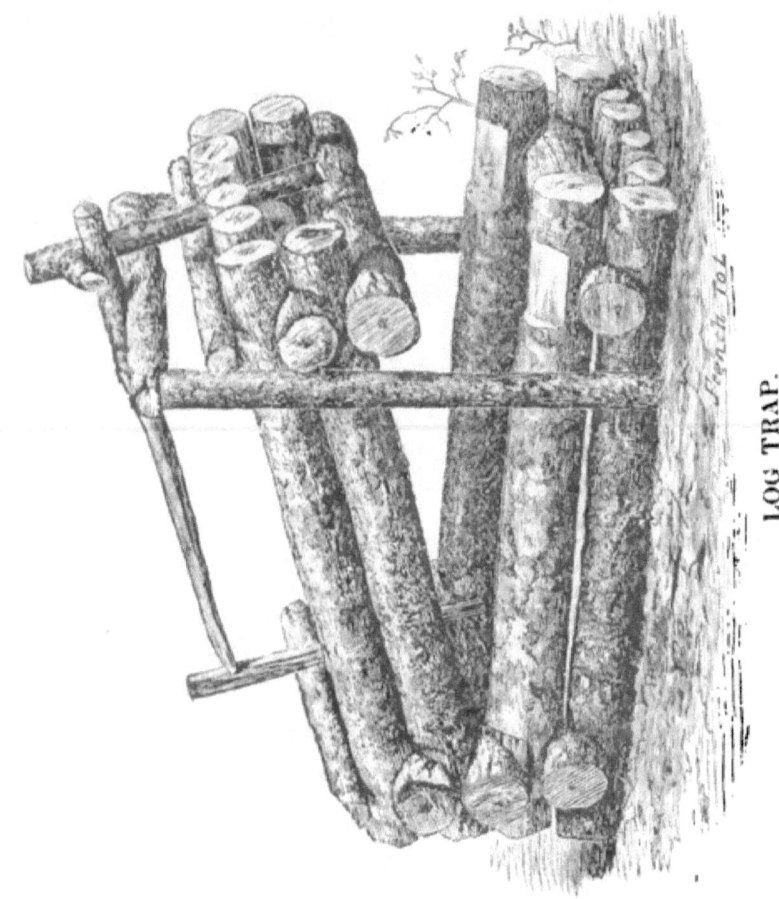

LOG TRAP.

her down. She stepped very accurately, and did not once stumble against any log.

All things considered, the log trap is the best for bear catching. It saves hard labor in carrying traps through the woods, obviates the trouble and fatigue often occasioned by hunting the lost bear and the lost trap. The log-trap is very seldom sprung by any other animal, and is usually a very certain catch for the bear. It is easily made; it lasts the hunter for several years, should he hunt on the same ground for so long a time; the bear is not afraid of it, and is not injured by it when entrapped. The bear is not afraid, as some suppose, of newly chopped timber. The trap should be made of hard wood. If tamarack, spruce, or hemlock is used, the logs should be ten inches in diameter; if it is made of birch, maple or hard wood, seven or eight inches is enough. The inside measurement should be six by two and a half feet; the height should be two feet. When baited and opened, ready for the entrance of his bearship, there should be an opening of about twenty inches. Two men can make one in four hours, and easier than they can carry a steel trap ten or twelve miles, as the case may be. The log trap, with its posts, cross-bar, lever, spindle, and hook are illustrated in the preceding cut.

Any kind of meat may be used for bait. Beaver's flesh, venison, and mud turtles are very good; all kinds of fish are also good. Fish-oil draws them well, and honey has a still better effect in luring them to a trap.

GRIZZLY BEAR.

Old Grizzly is the most savage of all his race. He is a native of the Rocky Mountains, where he roams at will, taking his march as far north as 61 deg., and southward to Mexico. He is to the American farmer what the Bengal Tiger is to Hindostan, or the lion to that of Central Africa. He is to be feared. Animals which we would not dare to approach fall prostrate before him, and will flee, as if for their lives, from the scent of one of them, whether it be alive or dead. But man is the monarch of all terrestrial creatures, and even this one is not without an innate dread of humanity.

Wood, in his "Mammalia," says that "A man who was engaged in duck-shooting, and whose gun was only loaded with shot, was suddenly alarmed at seeing a Grizzly Bear cantering toward him, having clearly made up his mind to attack him. For a moment the old man was in despair, but his presence of mind soon returned, and he made his escape in a very ingenious manner. Plucking some of the light fibers from his rough coat, he threw them in the air, in order to ascertain the direction of the wind, and then moved to one side, so as to cause the wind to blow from himself towards the advancing foe. As soon as the bear perceived the strange scent it stopped, sat upon its hind legs, wavered, and finally made off, leaving its intended prey master of the field."

He is morose, indomitable, and very tenacious of life. When wounded he becomes a dreadful antagonist. Among the native tribes that dwell in the northern portions of America, the possession of a necklace formed from the claws of the grizzly bear is considered an enviable mark of distinction. No one is permitted to wear such an ornament but the slayer of the animal. "So largely is this mark of distinction prized that the Indian who has achieved such dignity can hardly be induced to part with his valued ornament by any remuneration that can be offered."

The older males often come from their burrows in the winter in search of food; all the others hybernate through the season. The bears of this species are expert climbers, though many have supposed that this was not the case. It is said that men have been saved from the death-dealing clutches of the grizzly bear, when resistance or escape would be futile, by feigning death. This bear is addicted to the habit of burying his prey. He will dig deep holes for his victim, and cover it with grass or leaves, and go off satisfied with his work, and will not at once return to the spot. Men have suffered themselves to be buried by one of these rough creatures, and have afterwards escaped in safety.

Lewis and Clark, who discovered this bear, give the measurement of one of them as nine feet from the nose to the tail. His claws are wonderfully strong and very sharp; they are sharp on the edges, like a chisel, and thus serve the purpose of a knife; but they are not sharp at the point to serve as an awl, or other instrument for boring. They measure about five inches in length; they are quite crooked. The paws are no less a wonder for power and size. The fore paws are strong enough to tear to pieces a very large animal, and the hind paws to squeeze the same to death. The fore foot of a full grown adult is nine inches long; the hind foot measures eighteen inches, the claws included; the breadth of the hind foot is seven inches. The tail is short, and lost in the shaggy hair. The line of its forehead and muzzle is straighter than in any other species. The average weight of the grizzly bear is eight hundred pounds. The fur is very long and shaggy, but upon the young bear is very soft. The color of the cub is brown; with advancing age it assumes a mottled appearance. Whether the fur beneath the surface is brown or grey, it is so evenly tipped with white that it always assumes the grizzled appearance which gives the animal its name.

RACOON.

This animal is classed with the bears. It is strictly American, and ranges from Canada to Paraguay. The common racoon of the United States extends as far south as Texas. It equals the fox in size, and resembles it in the sharpened muzzle, and in the general shape of the head. Its nose tapers beyond the muzzle, and is flexible. Its sense of smell is wonderfully acute, being by all odds its most perfectly developed sense. It is nocturnal in its habits, and is not able to see well by daylight. Many of the tamed creatures become blind by the formation of cataracts upon the eyes. Though it is plantigrade, it is only when at rest that the foot is placed flat upon the ground. The claws are sharp. Its body is about two feet long, its tail one foot. The tail is bushy. The general color of the animal is a brownish or blackish grey; the tail has several rings of black upon it; the muzzle is of a dirty white, a dark brown mark streaks across the eyes, and one upon the nose runs high up on the forehead; the under parts are the lightest. It feeds in the summer upon small animals, birds, insects, and eggs, but in general prefers a vegetable diet. It eats nuts, juicy fruits, and vegetables; it frequently makes havoc upon poultry-yards. It is an excellent swimmer, and is especially fond of fish, and of oysters and crabs. The tropical species of the racoon is, from its crab-eating propensity, called the crab-eater. Its sense of touch is acute, and is of great service in opening the shells of oysters and crabs. The racoon has the habit of dipping its food into water before eating it, and is, therefore, sometimes called the Lotor or Washer. It sits upon its haunches while feeding, and holds its food pressed between the paws, but not grasped in either paw. The racoon is easily tamed, and has been the pet of so many that its habits are quite well known. It is inquisitive, lively, capricious, and, when tamed, tenaciously remembers an insult. The borders of the sea, and the margins of swamps and river banks are its favorite localities. It spends much of its time in the hollows of trees. In the winter it hybernates. The mother rears her young in one of these hollows, the entrance to which is generally high from the ground. It breeds in the spring; the litter numbers from four to six.

Coons are usually hunted by the dog chase. This method is generally productive of great fun to the sportsman. When the coon has been "treed," a fire is built underneath, and by the light of it the animal is found, and captured by any swift climber.

The hunter takes them by the steel trap. The traps are set in some of their paths, by the side of streams, or in cornfields, to which places they frequently resort. When put on the paths in the fields, the traps are in some way secreted, are covered with dirt, grasses or leaves. They are baited with fish; the skin of the codfish, when roasted, is the best bait. If the traps are put near water they may be furnished with the same sort of bait. Their relish for roasted fish more than matches their cunning, and they are likely to be taken. An excellent method of trapping them is by setting a trap in water, and putting a piece of bright tin on the pan of the trap. They, being very fond of shell fish, are very curious to know what the bright tin is: they seem to think it is the shell of some animal. There is no better method than the last one named.

XIII.

FELIDÆ, OR CAT TRIBE.

Here we find a tribe of fierce yet beautiful animals, among which stands one prominent in stern grandeur and majestic bearing, a most formidable foe, that King of beasts, the lion. His domain is now limited to Africa, and certain parts of Asia. Our American hunter, as such, therefore, cares nothing for him.

In the same group of animals we find our common domestic cat. She is not specially valuable to the hunter, nor dear to those who think there is no animal so worthy of admiration and of petting as the horse or dog. Poor pussy is almost always sadly calumniated by the latter class of people, who set at great disadvantage her poorer traits of character by the most admirable ones of the dog. But she is not without her friends, and in the lap of their care and appreciation we leave her, assured that there she, representing the one extreme of the feline race, is as safe from the sneers and provocations of her enemies as are we from the clutches of the representative of the other extreme, so long as Africa keeps her children at home, and circus-managers keep their strong birds caged.

As a medium between these two extremes, we find in this country the Lynx, Cougar, and Jaguar, animals more rapacious than the domestic cat, less voracious than the lion, larger than the former, but not the peer of the latter. Still, since they possess the family traits, they are the most formidable, most blood-thirsty, in short, the most typical of the carnivora. They doubtless fall more victims than other carnivora, but they also reject more of these for food, often refusing to eat carrion, while other animals less voracious, but more disgusting, prefer the most putrescent flesh. "Their frame is vigorous, but agile, their limbs are short, the joints are well knit but supple, and every motion is easy, free, and graceful. They leap and bound with astonishing velocity. Their footfall is silent, the foot being provided with elastic pads, namely, a large basal ball or cushion, and one under each toe. The claws are of enormous size, hooked and sharp." They are wonderfully constructed. As the felidæ are digitigrade, and at the same time great climbers, and as they seize their prey by pouncing upon it, some provision must be made for the animal to walk upon the toes without injury to the claws, which must be kept sharp. "By a beautiful structural conformation of the bones, ligaments and muscular parts, they are always preserved without effort from coming in contact with the ground, and are retracted within a sheath, so as to be kept sharp and ready for service. This involuntary retraction, counteracted only by the action of muscles, is affected by two elastic ligaments, so contrived as to roll back the ultimate phalanx which the claw encases, and bring it down by the outer side of the penultimate phalanx which is flattened off to remove every obstruction. From this position the talon can be thrown forward in a moment, the action of the double elastic spring being counteracted by that of the flexor muscles. In the act of striking with great violence, the flexor muscles strongly contract, brace up the tendon, and throw out the talon, which, when the act is over, returns to its sheath."

The structure of the teeth is also strongly indicative

of their destructive energy. The four canine teeth are admirably constructed for tearing the flesh, and they do tear their meat; the molars are as well suited to the pecking bites which they give to the torn pieces of flesh. The surface of the tongue is scarcely less remarkable. Its entire surface is covered with conical projections, which point towards the throat. These projections, or large papilla, make of the tongue an excellent instrument for cleaning off the small particles of flesh which would otherwise be left upon the bones which the animal may be picking. The skulls of this family bear marked resemblances in the various species.

The senses are in high perfection, with the exception of the sense of taste, and it is also true that the sense of smell is less acutely developed than in many other animal tribes. The sight, hearing, and feeling are all very acute, the first being adapted for nocturnal, as well as diurnal vision, the last of these three senses having the singular but admirable provision of delicate feelers, in the whiskers. What better combination of provisions could be made for such sanguinary animals than these feelers, which warn of obstructions before a noise has been made by the body, in conjunction with the soft padded feet, which make no noise when the path has been determined upon.

CANADA LYNX.

This species of the Lyncine group which is found in the New World, receives its common name from one of the countries which it inhabits. The French Canadians, however, term it indifferently LeChat or Peeshoo. It lives in North America, in the region extending from the great lakes to the northern limits of the woods, and

THE CANADA LYNX.

is found in the Mississippi Valley, but rarely on the sea coasts. It is the largest American species of this group. It is as large as a setter dog, or intermediate in size between a fox and a wolf. It measures three feet to the base of the tail, which is about four inches long, measured to the tips of the long hairs. The form of the head is very much like that of the domestic cat; the nose is blunt; the head is round, and is about the length of the tail. Its general form is less elongated than the most of the family; the body is elevated at the haunches. The ears are pointed, erect, tipped with pencils of coarse, black hairs.

Its general color is grey above; the hairs are generally tipped with white, which accounts for the apparent changes in color of the same species at different times. In some specimens there is an indistinct mottling, but not often any defined markings. On the under parts it is lighter than on the back, and is sometimes white. The tail is greyish, but tipped with black. The fur is close and fine, shorter on the back than on the under parts. Its winter fur is long and silky. The same individual doubtless undergoes a change in the color and length of fur at different times of the year. Being an inhabitant of a colder climate than some of its kindred, it is more densely clothed with long fur than they.

Its average weight is twenty pounds. It walks the lightest and stillest of any animal. It will walk on new fallen snow and not sink more than six inches. Its track is larger than that of the Cougar or American Panther; because its toes, which are heavily furred, are spread far apart when the animal bears its weight upon them. Some will make a track nine inches long, almost as long as that made by the black bear.

It generally lives in the deepest woods, rarely approaching the habitations of men. It is most abundant in its most northern homes. It is strong and active, an excellent climber, and a good swimmer. Its gait is by bounds straightforward, with the back a little arched, and lighting on all four feet almost simultaneously. It feeds upon small quadrupeds, birds, fish, and is very fond of grouse. It will not attack any of the domestic animals unless driven by hunger. It is quite sullen and suspicious, and is not easily tamed. It is not courageous, seldom attacking large quadrupeds. It is not as prolific as the most of the cat tribe; it breeds but once a year, and the number of its young rarely exceeds two. The mother secretes her young until they are large enough to follow her. Its fur is of some value to the hunter; its skin is an important article in commerce. It can be taken like others of the carnivora. If the steel trap is used, it is to be concealed near one of its favorite haunts, and unless some of the flesh of an undevoured animal is close to it, it must be baited with some of its favorite meat. The scent of the beaver is excellent bait for the lynx, because it is especially fond of that. Some hunters use a pallet of birch bark to conceal the pan of the trap; upon this covering they put a piece of the skin of some animal, strongly scented with the castoreum of the beaver, or other favorite scent. The lynx, in its attempt to drag away the skin is caught in the trap. It is not necessary to use a large clog; as the lynx gives up easily, sometimes without an attempt to escape from the trap.

I frequently make a pen near the trunk of some tree: the sides of the pen I run out about eighteen inches from the tree, and have a space of about fifteen inches at the base of the pen, which is a V shape: close to the tree, or the apex of the pen, I put several pieces of bait,

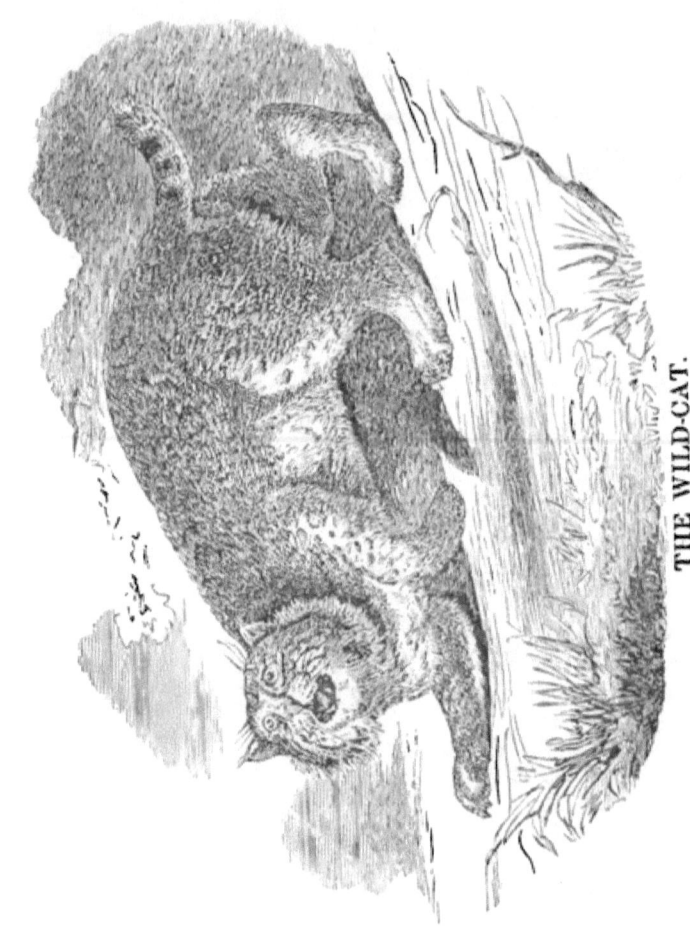

THE WILD-CAT.

and conceal the trap outside of it, near the base of the pen. The lynx will not eat the meat where it finds it, but will carry it off to eat; and, if it does not step on the trap when it goes for or away with the first piece of bait, it will doubtless be caught before all the pieces have been eaten.

THE WILD CAT, OR BAY LYNX.

There are no long-tailed wild cats in this country. Our American wild cat is lynx-like, and is known as the Bay Lynx. The true wild cat is found on the Eastern Continent. It was once very abundant in England, but is now nearly extinct. It became a great pest, and was hunted in the chase by the aristocratic leisurist, and driven before the hands of toil and honorable industry, as men reared cities and planted fields where this fierce creature had roamed at will. In color and contour the Bay Lynx bears many resemblances to the Canada Lynx. The throat is surrounded with a ruff of long hairs; the ears are black on the inside, and have a white patch near the tip; they are not tufted: the legs are long, and the hind feet are slightly webbed. The soles of the feet are naked. The fur is short and coarse. Its weight is from twenty to forty pounds. It mews and purrs like the domestic cat when it is in confinement. When roaming free, its caterwauling may be heard for a long distance. It feeds upon small quadrupeds, grouse, partridges, and other small birds. It frequents streams of water, where it feeds upon fish. The wild cat of the south frequents canebrakes and briery thickets in search of game. It is not cowardly, as some suppose, afraid to attack a quadruped larger than itself. It is exceedingly fierce, more so than the Canada Lynx, and will waylay deer and kill them. I have seen the tracks of a wild cat coming up onto the trail of a deer, and could tell by the tracks at what point it had jumped, and when it had touched the deer. The wild cat had dragged its hind feet upon the ground for a little ways, and had then jumped upon the back of the deer. I have known instances of this kind when the deer had run on about sixteen rods, with the ugly fiend tearing into its flesh, before yielding, the hair torn out and scattered along the path.

The lynx will come to almost any kind of bait, even flesh of its own sort. It is fond of venison. Like the rest of its tribe, it buries the flesh of its undevoured game. Traps for this animal are to be set in the same manner as for the Canada Lynx.

West of the Rocky Mountains there is a variety of the wild cat called the Red Cat. Its color, which is a rich chestnut brown on the back, gives to it its name.

THE COUGAR.

This carnivorous animal belongs to the cat tribe. Few creatures have been known by such a variety of names. Cougar is its proper and scientific name. It is extensively spread over North and South America, from Canada to Patagonia. It is more scarce than formerly, and its range is contracted. Advancing civilization will still further reduce its numbers and its hunting grounds. In South America it is generally called the Puma; in the United States west of the Rocky Mountains, it is called the Californian Lion, or the American Lion. For the name of lion the cougar is indebted to its uniform tawny color, neither of them being streaked or spotted, as are many of the felidæ, or cat tribe. East of the Rocky Mountains it is commonly called the Panther. The true panther, however, is confined to the Eastern Continent.

THE COUGAR

For the name of panther the cougar is indebted to its marked resemblance in habits to the pardine or spotted animals of the tribe. The cougar or puma loves to hide in the branches of trees, or to haunt grassy plains, where it can watch its prey, and pounce down or leap out upon it, and with its death-dealing paws make sure of its victim. There it destroys wild cattle in great numbers. In its northern home the cougar lives in the forests, or retreats to high rocky ledges, where it is literally lord of all it surveys. Hunters call them panther ledges.

In its general contour it is elegantly formed, although the head is quite small and the limbs very thick. The fur is thick and close. The general color is silvery-fawn above, fading into a handsome greyish white on the under parts. The base of the ears, sides of the muzzle, and end of the tail are black: the breast is almost a pure white, as are also the whiskers. The young are marked with three rows of blackish brown streaks; and the sides, shoulders and neck have clouded spots of the same color, thus resembling the leopard. These markings gradually fade, and when the animal has attained its full growth they have entirely disappeared. Only one member of the cat family existing on the Western Continent is larger than the cougar. Six and one-half feet is the total length of this animal, of which two feet is the measurement of the tail. Its eyes are large, grey, and look very fierce; its limbs are remarkably strong; the teeth are sharp, and the claws are long and heavy.

The animal, being a very expert climber, is quick and nimble in its motions. It will catch a piece of flesh from any animal at which it strikes. Like the other members of its tribe, it is light, stealthy, and silent of foot; the senses of hearing and smelling are acute, and it is swift of attack. It is nocturnal, as its stealthy, sneakish disposition might determine it to be. It has a special relish for the deer family, any of which it can capture. These it generally secures by pouncing upon them from some cliff or branch overhanging their paths or watering places. Its greed for them is almost insatiable, it being rarely satisfied with a single creature. It is also very fond of sheep and cattle. Hunters, therefore, kill them when they can. It is cowardly, and is oftener known to run from a man than to him. Instances are on record in which men and panthers have come unwittingly upon each other, and the latter have walked off completely mastered by the fixed gaze of the master human eye. When taken young the cougar is easily domesticated, when wounded by man is a very dangerous foe.

On the Pampas the puma is hunted with dogs, or on horseback, when the hunter uses the lasso and the bolos. In our own country it is almost always taken by rifle shot, while it is perched in some tree. It may be taken in the steel trap. The trap, after being placed near the remains of some animal which it has partially devoured, must be secreted. Its voracious appetite forbids it to leave any portion of the carcass undevoured, and it will frequent the spot until this is accomplished. It covers the food which it wishes to leave for a time.

XIV.

THE FOX.

The fox is a well-known carnivorous animal, belonging to the vulpine division of the family canidæ. It may be readily distinguished from the dog, the wolf, the jackal, etc., by the construction of the pupil of the eye, which is vertical, and not, as in the latter cases, circular. It may also be distinguished from the dog and the wolf by its sharp pointed muzzle, long, bushy and cylindrical tail, lower stature, slender limbs, and short, triangular ears. It is nocturnal in its habits, while they are diurnal. The dentition is the same as that of the wolf.

The voice of the fox is a sort of yelp, though varying in its tones. Its color is a reddish brown, intermixed with black and white hairs. The tail, frequently called the brush, is colored like the rest of the body, except at the tip, which is white. The fur is very long, thick, fine and glossy: it has about the warmest fur of any animal. It never seems to suffer from the cold, although it is its habit to get upon upturned roots, knolls, and other high places, the better to watch everything around it. The fur of the winter-killed fox is, because of its warmth, very valuable.

The fox breeds once a year, having from four to eight young ones at a litter. The entire care of the young devolves upon the mother. She prepares for them a bed of dried leaves, grass, and mosses. The kittens are very playful, playing like the kittens of the domestic cat.

They remain about four months under the immediate care of the mother fox; she is very resolute in their defense.

The fox is solitary in its habits, living alone in a burrow, which it has usurped, or has made by digging up earth, if possible around and beneath the roots of a large tree. Sometimes these burrows are made between huge stones, or in other secluded situations. They are generally in the neighborhood of a rabbit-warren, preserves of game, or farms. It is a very unsocial creature, and cannot well be domesticated. Adults are ferocious when placed in confinement, and soon die. Though slightly made it is vigorous, and bites with great severity.

It is a very rapid traveler, and in many of its motions is very graceful. It has a light, tripping gait, and puts one foot so nearly in front of the other that if a cord were stretched along its track, it would lie upon the tracks of all four feet at once.

There is a gland near the root of the tail which secretes a very fetid substance. The scent is so strong that a man can easily track the animal for some distance. The scent clings for a long time to any object which it has touched.

The fox feeds upon small quadrupeds, birds, fish, and the poultry and eggs which it delights to steal from the farmyard. Its taste is not at all fastidious, as it will eat small reptiles and carrion. We cannot say if stolen food is any sweeter to the fox for being stolen; but it is very thievish, and sly in its attempts to get its food; it steals under cover of night, and not satisfied with this dark covering, its manner of advancing is stealthy, pouncing at the last upon its prey. It is fond of fruit, especially of grapes, the sweet ones. Old Reynard doesn't like " sour grapes." The fox is still further responsible for our vocabulary of foxy phrases. The epithet " foxy," because of the characteristics of this little animal, is proverbially applied to the cunning, tricky, and unscrupulous knave. Sly as a fox, or cunning as a fox, are alike characteristic expressions. Many anecdotes are extant illustrative of the cunning of the fox in pursuing prey, or in eluding its pursuer. It is not a reliable creature, any more than are some people; for it demands what it will not give; namely, a fair chance to attack whom it will, but giving little of the sort to those that would fall upon it. Foxes are said to have been observed approaching water-fowl by swimming slowly with a turf in their mouth, so as to remain concealed. It is also stated in Chamber's Encyclopedia that "a fox was seen to approach a group of hares that were feeding in a field, with a slow, limping motion, and having its head down as if eating clover, till it was near enough by a sudden rush to secure a very different food." This is entirely credible, for it is a shrewd mimic; it imitates very acurately the motions of other animals, to avoid attention. Foxes do not herd together when they hunt, but go singly, each one playing its own tricks, getting and eating its own food. It knows that some animals are afraid of its scent, and by taking advantage of this fear, sometimes secures to itself desirable food by getting in ahead of them, knowing that they will not come near the haunt again. Wood, in his "Mammalia," gives an account of some nefarious attempts of a tame fox to steal the food intended for the cats and dogs.

Naturalists recognize fourteen species, which are distributed through all latitudes: they are found in most parts of Europe, extending into Northern Asia, in Africa, and in America. Six of the fourteen species are found in the United States; namely, the Red, Cross, Silver, Grey, Prairie or Swift, and the Kit Fox.

The common Red Fox resembles the common fox of Europe; its fur is, however, longer, finer, and is of a bright rufous brown, and is much more valuable, forming an important article of export. It is often used as an article of trade. About eight thousand are annually exported to England from the fur countries where they are numerous. It has less vigor and endurance than the English fox. The American fox has great breadth of feet, giving to it greater capacity for traveling on the snow; it has long hair, covering the back part of the cheeks, a short muzzle, short legs, and a very bushy brush. The Red Fox varies in its color more than some distinct species, some are dark, almost black; some are of a pale yellow; while others are of a reddish fawn. It measures three feet from the nose to the tip of the tail, and weighs from nine to fifteen pounds. It is a great rover, traveling from ten to twelve miles from home in a night. It is quite irregular in its tramps from home, but very regular in going to feed, even upon the hunter's bait when it has found it. It will frequently go for miles in precisely the same tracks, and night after night will follow them up. It is so keen scented that it has been known to leave its track for some rods to catch a mouse. Its tracks show that it has cautiously approached within a few feet of the mouse, and has then pounced upon it. It is said to have a habit of leaping along upon the snow to scare out the field mice. It possesses as much cunning and craft as belongs to any member of the fox family. It is found from Pennsylvania to Canada, and from the Atlantic to the Missouri river.

The Cross Fox is similar to the Red Fox, but is distinguished from it by having a longitudinal, dark band along the back, crossed by a transverse band over the shoulders. It is quite easily caught in its burrow, or driven out of it.

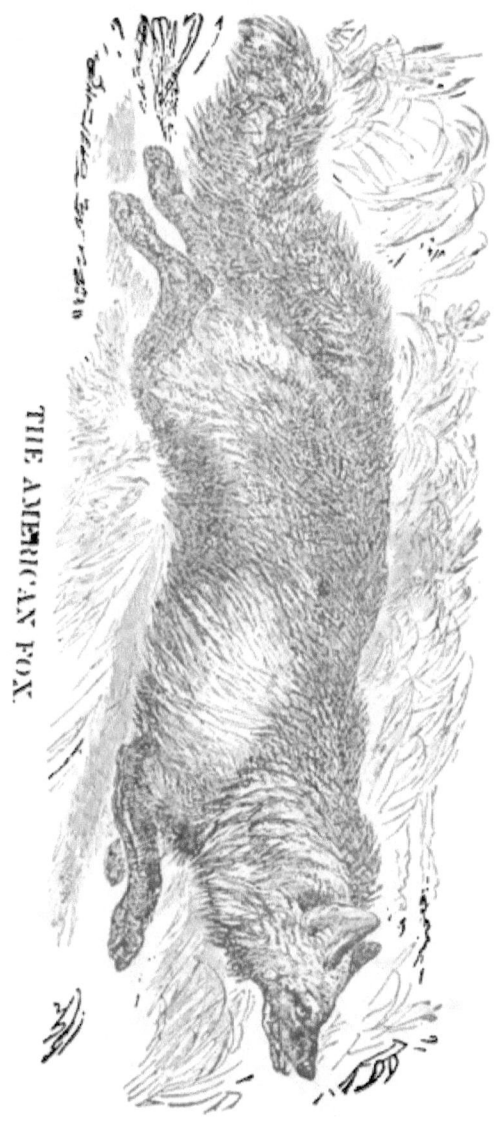

THE AMERICAN FOX.

It is found in a region which is narrow, but which stretches through northern New York, Canada, Michigan and Wisconsin.

The Grey Fox is very abundant in the southern states: it is occasionally found as far north as Canada. It is about the size of the Red Fox, is very cunning and sagacious. Its color, indicated by its name, is light on the upper side, and yellowish on the under side: its ears and feet are black. The fur is short, coarse, and thin. Its tail is somewhat flattened. In short, it is a cheap kind of fox.

The fox hunt, from its exciting nature, the cool calculation and stratagem requisite, has long been practiced, and known as the "King of British National sports." It is practiced in the southern states of our own country. The fox will sometimes go forty or fifty miles before the hound without stopping. To the practical hunter, and especially to the woodsman, this method of capturing the fox is not at all practicable. It is supposed by many to be effort thrown away to attempt to catch the fox by the steel trap; it is, nevertheless, done with great success.

There are different methods of setting the trap, as different localities demand. One very good method is as follows: put a piece of meat in a spring that does not freeze over in winter; put it in so far from the dry ground that the fox cannot reach it without getting its feet wet, for, like cats, they do not like to get their feet wet; put a stone in the water between the water's edge and the bait. The stone should be large enough to come just above the surface of the water, and a piece of moss should be put on it for the fox to step its foot upon. The moss will not be as apt to awaken any suspicion as the stone would. The fox having once stepped upon the moss-covered stone, in search of the food, will be very sure to come again, and so regular is it in its habits that the hunter can tell about when it will come. Let him now remove the stone and put in its place the steel trap, covering it with mud, and putting the moss on the pan of the trap. By this method the hunter will not fail to get his fox.

I once caught one which was very lean; the skin was also very poor. The trap caught it very low on the hind legs, and in an effort to extricate itself, it had torn the flesh very badly upon the jaws of the trap. I would not keep it, but let it go, saying that I would catch it another time. After about six weeks, during which time I had not seen my fox, I caught it again in a trap. I knew it to be the same fox; for the broken bone in healing had left a bunch on the leg, and the wound above the fracture was not entirely healed over.

Again, the trap may be set on a hummock, sand-bank or other slight elevation, as foxes delight in resorting to such places to look about them. The trap should be concealed about two inches under the ground: the meat which is used for bait should be put over the trap, but not on the pan, and scattered all about it from six to eight inches around: the ground must be nicely smoothed over, or left to look as natural as possible; for the fox is very suspicious of change: some bits of meat may be scattered on the ground some little distance from the trap, to attract attention; they may be scattered along one, two, or three rods off: if the surface was very smooth it should be brushed with a quill or bush: all footsteps should be carefully erased. This method should be pursued in the fall before snow falls, or the ground is frozen. The fox in digging for the meat will find the trap.

In Allegany county, New York, I once set a trap and covered it with chaff. I found the bait had been eaten and the trap tipped over. This performance was repeated several times. I then set two traps very near each other, one larger than the other. I found the little trap sprung, and the fox in the larger trap. It had avoided the larger one; it was caught in the small one; but being too small for it, it had extricated itself, and in the effort had inadvertantly sprung the large trap. The saying that "It takes a rogue to catch a rogue" is not without some verification.

I have known many persons who could not get a fox to come close to a trap. A fox is suspicious of a man's track, and will not follow to the end of it. If the hunter will, after setting his trap, walk on in a circuitous course, in a short time coming back upon the track at a place walked over before the trap was set, he will surely catch his fox. I use this method in the winter, when the snow is on the ground and the animal can see the track. Since the ground is at this time frozen, I cover the trap with ashes. I do not believe that the fox is afraid of the scent of iron.

It is better not to wash the trap after having caught a fox, as the fox scent is of itself excellent bait. Meat of various kinds makes good bait: the musk of the muskrat is very good. The scraps left from the trying of lard is the very best for baiting purposes. The fox is so fond of it that it will follow one a long distance, eating every particle that the man may have dropped, even though the pieces should be no larger than a pea. Though the fox is so exceedingly cunning, the hunter's knowledge of its favorite food often proves too much for the little creature.

There is a fox known by hunters as the Samson Fox. There are but very few long, coarse hairs in the fur, and those look as if they had been crisped. They give to the animal a dark, dull color. The skin is almost worthless. It is in other respects like the common Red Fox.

XV.

THE WOLF.

"Few animals," says Wood, "have earned so widely popular, or so little enviable a fame as the wolves. Whether in the annals of history, in fiction, in poetry, or even in the less honored, but hardly less important literature of nursery fables, the wolf holds a prominent position among animals.

There are several species of wolf, each of which species is divided into three or four varieties, which seem to be tolerably permanent, and by many observers are thought to be sufficiently marked to be considered as separate species. However, as even the members of the same litter partake of several minor varieties in form and color, it is very possible that the so-called species may be nothing more than very distinctly marked varieties.

These voracious and dangerous animals are found in almost every quarter of the globe; whether the country which they infest is heated by the beams of the tropical sun, or frozen by the lengthened winter of the northern regions. Mountain and plain, forest and field, jungle and prairie, are equally infested with wolves, which possess the power of finding nourishment from their united bands in localities where even a single predaceous animal might be perplexed to gain a livelihood."

The wolf belongs to the dog family, and bearing many marked family traits, needs no detailed description. Ferocity, cowardice, and cunning, characteristics attributable to wild beasts, are the natural developments of this species of the canidae tribe, the Canis Lupus. In structural formation it is very similar to many species of of the domestic dog, but yet would not be mistaken for it, while savage brute is depicted upon its countenance, and wildness is stamped upon its whole being.

In North America there are two well-marked wolf sections. In the smaller section is found the Prairie Wolf. Its skull is slender, and its muzzle is fox-like, and the crests are not prominent. In the larger section the skulls are higher, the muzzle is more blunt, and the cranial crests are well developed.

The common Grey or Timber Wolf is found in the northern part of North America. It measures from three to four and a half feet in length; the tail is from seventeen to twenty inches long. Its color is greyish above, the hairs are tipped with black, giving to the animal a mottled appearance; the under parts are much lighter.

The Indians take this wolf by advancing upon it by taking first a circuit of several miles about it, and following this by a series of narrower and continually narrower circles, until they have come within shooting distance of it. Large premiums have been awarded for the destruction of this creature. This species of wolf is the only one found in Canada. It is stouter than the common European wolf: its hair is longer and finer.

The wolf on the upper waters of the Missouri is white; in the north-western states it is dusky, in the south black, and in Texas reddish. Except the antelope, it is the fleetest animal in the Missouri region.

Wolves were formerly much larger throughout the United States than now. A few of these giant wolves are still found in the most densely wooded mountain regions of the New England states, and in unsettled portions of the country which deer inhabit.

The Buffalo Wolf is the largest American species. In the far west these follow the buffalo in packs, and fall upon the sick or feeble ones of the herd, or upon the stragglers; but they do not when alone attack the vigorous ones. When in packs, and very furious, they will make an assault upon man or beast, and will generally devour them. A pack of wolves, when infuriated, will fall upon its prey with the greatest violence; but if any one of the pack is left alone, it will, like the sneak and coward that it is, howl and snap its teeth, but will not dare to advance. Some have supposed that these large wolves were a distinct species, and have therefore called them the Giant Wolf. It is probable that they are identical with the common Grey Wolf, but of larger growth; because in earlier days nature, having fewer to nourish, had given them a double portion. The color is due to climatic influences. The same is true of other animals. They are the largest, most robust, their native tendencies most perfectly developed in their most natural homes. Wild beasts are the most bestial in their wildest homes: they become enfeebled and intimidated when bounds are set to their rovings, by the building of towns, or by the woodsman's forest clearing.

The Rocky Mountain White Wolf and the Buffalo Wolf are the largest known animals of the Lupine division of the dog family. Next to these in size is the Timber Wolf. It is rarely seen in the day time; I have never seen but one in the day time, in its wild condition, unless it were in a trap.

The wolf is to be trapped in the same manner as the fox; only, the trapper must take pains not to leave any artificial work about the trap. Traps are sometimes set on paths which it follows in the swamps, or elsewhere. Everything must be left about a trap to look perfectly natural. They are great ramblers. After having killed a deer, they will often leave for days what remains from their first meal. They will sometimes be gone ten or twelve days, but are sure to come back to the same spot: they will not abandon it so long as any of the carcass is left. Such a place is, therefore, a very good one for trap setting.

XVI.

THE OPOSSUM.

The marsupials to which order the opossum belongs are confined to Australia and its adjacent islands, and to America. Fossils are still found here; but there is now only one existing genus found in this country. The distinguishing characteristic of these animals, and that which gives to them their name, is a pouch or marsupium in which the immature young are carried, and where they are nourished until they are able to begin to care for themselves. In some cases the pouch is only rudimentary, being but folds of the skin, and in which cases the young cannot be carried about in them. The Virginian Opossum is very prolific, often producing fifteen or sixteen at a birth. The mother conceals herself in some nest which she has made of dry grasses, and when her young are fifteen days old she transfers them to the pouch, where she keeps them for thirty-five days. When they are placed in the pouch they are less than an inch in length, inclusive of the tail. They have at this time no hair upon the bodies, are blind and also deaf. When they are fifty days old they are about the size of a mouse, have hair upon their bodies, and being then released from the pouch, the eyes are opened. The tails exhibit at this early period their prehensile character. The exact time when their deaf ears are unstopped is not known. A great amount of careful observation has been required

THE VIRGINIAN OPOSSUM.

to ascertain these facts, some of which Wood has given in his "Mammalia."

The opossums of America are especially remarkable for their robust forms, and their peculiar habits. There are about twenty species, of which the Virginian Opossum ranks first in interest to us. It is not confined, as its name might imply, to Virginia, but is found in many parts of the United States. In size it is equal to that of an ordinary cat, or, it is about twenty-two inches in the length of the head and body, measured over the curve of the back. The tail is fifteen inches long, is prehensile, but capable of involution only on the under side, and is covered with scales, through the interstices of which a few short hairs protrude. The inner toe of the right foot is converted into a thumb, destitute of a claw, and has a nail like the thumb of a child; it is opposable to the other toes, thus enabling it to grasp the branch of a tree with considerable force. The structure of the tail and feet is such as to make it an admirable climber. It is a five-toed plantigrade, the limbs are short, the claws are long and sharp. The teeth are numerous, amounting to about fifty. The head is long and pointed, the profile straight. The eyes are high on the forehead, are small, dark, prominent, and have a nictitating membrane, but no eyelids. The ears are large, thin, naked, and rounded. The tongue is rough. The snout is long, the muzzle pointed, naked, and moist. The nostrils are lateral. The mouth is wide, and the whiskers are stiff. The expression of the physiognomy is peculiar and unpleasant. The fur of the opossum is fine and wooly, and is long and soft, except on the head and some of the upper parts, where it is short and coarse. The general color of the fur is a dirty white, slightly tinged with yellow, and diversified by occasional long hairs that are white toward their base, but of a brownish hue towards their points. The ears are black, margined at the tip with white. The scaly portion of the tail is white.

The opossum is slow in its movements. It is nocturnal and nonaquatic in its habits. It is a forest resident, spending its time during the day on the branches and in the hollows of trees, in a torpid state. It often hangs suspended from branches of trees by the tail, and by swinging its body will contrive to fling itself to the adjoining boughs. It often sleeps while thus suspended from the limb of a tree. At night it prowls about, liking best the bright, still nights, visiting rice fields, low, swampy places, or barnyards. Its good climbing abilities makes the latter place almost defenseless; its sense of smell is in high perfection, and its voracity and destructiveness are remarkable. Its bill of fare, well proportioned to the cravings of imperious hunger, consists of small quadrupeds, some of which it eats by the brood at a time, reptiles, birds, insects, eggs, also roots, fruits, nuts, and juicy stalks of plants. It is very fond of maize, and its fondness for persimmons is almost proverbial. It has been seen hanging from persimmon trees while gathering and eating the fruit. It is cunning in its stealthy quest for prey. It is neither gregarious in its habits, nor graceful in its gait. It is fond of lying on its back in the sun for hours together, and is then frequently found alone. Its gait on the ground is heavy and pacing. When irritated, the animal emits a very unpleasant odor, and it is in many respects displeasing in its habits and appearance.

The trick so extensively known as "playing possum" is shrewdly managed by this little creature, though it often suffers severe torture at the hands of its would-be-captors. It does not often put this ruse in practice, until all other means of escape have proved futile: when it does yield to feigning death, only an old "possum" hunter can always tell whether the wary creature is really dead, or whether it plays dead. Some one says, "If a cat has nine lives, this creature surely has nineteen; for if you break every bone in its body, mash its skull and leave it for dead, it may still survive, and you may find it after a time creeping away."

It is a stupid fellow, and utterly neglects its safety, but being blessed with a good appetite, may be easily caught in traps set in its paths, which may be baited with any of its favorite kinds of food. Very rude traps are often the effectual means of its capture.

XVII.

THE MUSKRAT.

The Rat is classed among the rodents, or gnawers. The rodents have no canine teeth. The molars have flat surfaces; the incisors are exactly adapted to gnawing, nibbling, and scraping. The incisors of other animals would soon wear out, if those animals were subjected to the same method of eating, and so would they wear out in the present case, but for a provisional arrangement by which the worn-out particles are being constantly renewed. When the growth of the teeth is provided for, one other thing must be done, and is done. The chisel-like edge of the incisors must not be allowed to become dull, else the animal might as well have no fully developed teeth. The enamel and the dentine of each tooth is much harder on the anterior surface of the tooth than on the posterior, and this layer forms the cutting edge, as does the layer of steel on softer metal, composing a common chisel. The back surfaces of the teeth are worn away much faster than the front, and the edge is preserved.

The rodents are extensively spread over the globe. They are represented by the mice, rats and squirrels. One must be quite ignorant to know nothing of their physical structure, or of their vexatious and destructive habits.

Musquash, Ondathra and Muskrat are names given to this species of the rat. It is a native of North America,

THE MUSQUASH, OR MUSKRAT.

where it is widely known. Its color is a dark umber-brown on the back, passing into a brownish-yellow on the under parts. The fur is composed of fine, silky hairs, with which longer, coarser hairs are intermingled, especially on the upper side.

Muskrats are aquatic in their habits and are excellent swimmers. The hind feet are slightly twisted; the inner edge is posterior to the outer, thus becoming a very good oar when the animal is swimming. The hind feet are webbed; fur covers the tops of them; the soles are naked; the edges are margined with bristly hairs. The fore-feet have a wart-like thumb and four toes; the hind feet have five toes. The tail is about two-thirds as long as the body, is scaly, and between the scales there are short hairs. About two-thirds of the length of the tail is rudder-shaped.

They are nocturnal in habits, omnivorous in appetite, and yet they live principally upon a vegetable diet, upon roots and tender shoots. They frequently feed upon the dead bodies of their own kind, and upon the lacerated bodies which they find struggling in some trap.

We quote from Audubon and Bachman the following on the habits of these animals: "Muskrats are very lively, playful animals when in their proper element, the water; and many of them may be occasionally seen disporting themselves on a calm night in some mill-pond or deep sequestered pool, crossing and recrossing in every direction, leaving long ripples in the water behind them, while others stand for a few moments on little hurdles or tufts of grass, or on stones or logs, on which they can get a footing above the water, or on the banks of the pond, and then plunge one after the other into the water. At times one is seen lying perfectly still on the surface of the pond or stream, with its body widely spread out, and as flat as can be. Suddenly it gives the water a smart slap with its tail, in the manner of the beaver, and disappears beneath the surface instantaneously, going down head foremost, and reminding one of the quickness and ease with which some species of ducks and grebes dive when shot at. At the distance of ten or twenty yards, the Muskrat comes to the surface again, and perhaps joins its companions in their sports; at the same time others are feeding on their grassy banks, dragging off the roots of various kinds of plants, or digging underneath the edge of the bank. These animals seem to form a little community of social, playful creatures, who only require to be unmolested in order to be happy. Should you fire off a fowling-piece while the Muskrats are thus occupied, a terrible fright and dispersion ensues; dozens dive at the sound of the gun, or disappear in their holes."

Its burrows are upon the banks of streams. The entrances are under water, and lead up sufficiently high to be beyond the reach of the greatest freshets. They are made of mud, in a conical form; long grasses are intermixed with the mud; dried grasses are used for beds within the burrows. The winter burrows are about fifty feet from the water; by water communications with the burrows the entrances are still from the under side. In swampy localities they build their houses of sticks, grasses, reeds and small branches of trees. They are in other respects like the burrows upon dry ground. When the surface of the water in which they are freezes over, they make feeding holes; these holes they protect from frost by a covering of grasses. They seem to prefer swamps to deep, clear water, and probably because they find their food, grasses, pond-lilies, etc., more readily in such places. They swim near the surface of the water and can easily be seen through the ice.

Hunters spear them through the walls of their burrows. The spear is a piece of iron about three feet long, a half inch bar of iron; the point is sharp and has two beards. Great caution should be used in approaching their burrows, as they are suspicious of a strange noise, and will take to the water for safety. The hunter should carry two spears; so that when an animal is caught the other will be ready to follow up those that run out. They are also speared through the ice. The hunter should wear moccasins or something very soft, in order to approach noiselessly. In striking a spear through the ice it is not uncommon to catch several at a time. An experienced hunter can tell by the condition of the house where the animal lives, and also the location of the chambers of the burrows, and its probable position in the chamber. They are also taken by the hunter's spear at their feeding places; these are the surest places for this mode of capture. After swimming some distance under the ice they stop, and putting the head close to the ice they exhale; the air which they emit from the lungs remains in bubbles until it has had time to become oxydized by the water, when they inhale the purified air and swim on. Taking advantage of this, the hunter can spear them in great numbers; he can see the bubbles through the ice, and has time to throw his spear while the animal is stopping to catch its breath. The Indians spear them.

The steel trap makes a better instrument for their capture than the spear. Traps should be set at the entrances to their burrows, and in the principal feeding places. They should be so set that the rats in endeavoring to extricate themselves shall be flung into deep water.

The best seasons for rat spearing or trapping are the autumn before the snow has fallen, and the spring after it has disappeared. In the summer these animals burrow in the banks of lakes, or streams, or in their houses, and they make canals sometimes many yards in extent. In the autumn when they are attacked they retreat to these canals.

The flesh of the muskrat and the musk of the animal are the best bait. It is best when bating traps set at a burrow entrance to put the bait about six inches beyond the trap on the land side; because the water is the natural home of these rats, and they will therefore oftener approach from the water side of the trap. The bait should be set in very shallow water; or if the water is deep close to the land, put it upon the land and the trap very close to the shore. When the musk is used for bait, it is to be put upon a piece of wood, or a twig, and stuck into the ground. Muskrats frequently follow each other for long distances on account of the scent which they emit. Their odor is offensive, yet less so than that of the mink. They are not very cunning, and may be taken in any steel-trap or in an ordinary box-trap. Musk, which is to be used in bating them, can be obtained the best in the spring. It should be preserved in alcohol. They are the prey of a great many animals.

They are prolific, breeding twice a year, and producing from four to ten at a time. The nests in which they breed their young are in a chamber at the extremity of some canal or road.

THE BEAVER

XVIII.

THE BEAVER.

The Beaver belongs to the order Rodentia, the distinguishing characteristics of which have already been noticed. One South American rodent is larger than the Beaver; with this exception it is the largest among existing species of its order. The American fossil Beaver is the largest rodent, either among fossil or living specimens. It is said to be contemporaneous with the mastodon, and is therefore one of the oldest of the mammalia.

The home of the American Beaver was more widely spread over the continent than that of any other animal. The boundary lines of the homestead of this veteran were by no means stingily set, extending as they did over and beyond the limits of the United States. The Arctic Ocean marked its northern boundary, the Atlantic its eastern, and the Pacific its western. Being aquatic in its habits, its resorts were upon the streams and the smaller lakes.

The Hudson's Bay territory, the Lake Superior regions, the head waters and the tributaries of the most of the large rivers were the favorite resorts, and the places where the beaver congregated in the greatest numbers. Advancing civilization, like the utilitarian that it is, has driven it beyond the thickly settled portions of the country. Its homes in the more sparsely settled sections, or the wilds of the far West, are the same as formerly. For the trapper who would understand the habits of the beaver to increase the profits of his traffic, for the scientist who would study their ways and means of living for the knowledge he would acquire and could impart, there is no better place on the continent than that known as the Lake Superior region. There they are very abundant and their works are numerous, large, and very perfect.

In its physical structure the American Beaver is of comparative low order. Its general contour is squirrel-like, or rat, or rabbit-like, being of the same order as these animals. Its form is quite clumsy, the body is the largest through the middle and tapers gradually towards the extremities, thus making its upper margin, viewed from the side, very plain. The ears are exceedingly small; they are short and obtusely pointed. The sense of hearing is acute, the internal ear being relatively large. The eyes are small and devoid of any marked expression; they are set midway between the nostrils and the ears. The vision is short. The sense of smell is acute.

The jaws are furnished with the most perfect rodential teeth. Unlike the rodents as an order, the beaver has a free horizontal movement of the inferior jaw, a motion made from side to side. This increases the power of mastication. The three animals of this order previously referred to are also exceptions to the rule. The muscle which provides for this horizontal motion of the jaw, the wonderful perfection of chisel-like teeth furnishes another instance of the adaptation of wants and means. There are twenty teeth, eight in each jaw are molars and two are incisors. These incisors, or canine teeth as some prefer to call them, are rootless teeth; half of the molars are deciduous; in the perfected state the molars are rooted teeth.

The legs of the beaver are built in conformity to its amphibious nature. The fore legs are short, being about five inches long, inclusive of the claws. There are five claws. The hind legs are long and the claws are larger than upon the fore feet or hands. The feet are completely webbed. The palms of the hands and the soles of the feet are padded and naked.

The tail, which is composed mostly of dense fatty tissue, is broad and thin, and is slightly convex on both surfaces; the edges are very thin. It is attached to the body by a posterior projection measuring about four inches in length; in this projection are found the strong muscles which control the motions of the tail. In its outline the tail is a good oval except at the anterior extremity where it measures about three and a half inches in width. The tail is covered with a lustrous, black, scale-like substance; this substance, however, is not horny, but is a thick dermis so serrated as to present the appearance of perfect scales. It is continuous upon both surfaces of the tail. The dimensions of a beaver's tail average six inches in width, and twelve in length; the thickness through the middle of it is one inch.

The fur on the back is nearly two inches long; on the extreme under parts the fur, which is destitute of the long hairs, is very fine. In its northern homes the color of the beaver is black, or a chestnut brown. The under fur is darker than the outer fur or the long hair. Some specimens in the Lake Superior region are coal black. In California they are often of a leadish color. The average size of the beaver's body is two and one-half feet exclusive of the tail. Its weight is forty-five pounds.

The beaver is a burrowing animal; its claws are therefore constructed for excavators. The hands and the arms, or the fore legs, are powerless in the water, and are laid back upon the body when the animal is swimming, by this means not only taking themselves out of the way when of no service, but by the act facilitating the motions of the body. When working upon the land they become excellent implements for burrow digging, and for other purposes in the construction of its works.

When the beaver is cutting wood it sits in an upright position, the tail and the hind legs giving it a firm support. When frightened it sits in this position, and depends upon its acute hearing to learn of threatening danger. When in the water the legs and tail become of the most service in the act of swimming. The legs are thrown horizontally backwards; the head is just out of water; the hinder parts of the body are sunk in the water. When they wish to swim very rapidly they go to the bottom. When in very deep and still water they will, by the motion of the tail, which they use as a propelling power, make a swell upon the surface a foot high. The tail is the rudder, and, therefore, determines the position of the head, and the direction of the body. Like the muskrat, it swims for long distances under water without breathing. When the animal is frightened it slaps its tail upon the water, and always dives immediately afterwards. By this means it makes a very loud noise, loud enough to alarm any beaver at a long distance. Another function is attributed to the tail of the beaver. It is sometimes called a trowel, because it is used in packing dirt upon its dams and its burrows.

When the animal walks it raises the body only just above the ground; the tail drags upon the ground. Its gait is slow and shuffling. It is not by rapid strokes, but by exceedingly skillful ones that it fells forests and builds extensive dams. It is a graceful and rapid swimmer. The water is its natural home. It visits the land to obtain its supplies of food, and the materials for the building of its homes.

The beaver, like other rodents, is a vegetarian. It is fond of coarse grasses that grow in the margins of its ponds, and in its meadows; it also feeds upon the roots of the pond lily, and of other plants. It is extremely fond of blueberries and raspberries. In the summer it frequently cuts off all the raspberry brush it can find. It is generally known to be a great wood-chopper, and it is also believed by many that it will fell trees to get the bark and clear wood for food. It does not feed upon either in the summer. No fresh cuttings of large trees are found from early spring until the fall, at such times as it begins its cutting for the winter. Sometimes its food gets sour from long soaking, and its winter supply is not sufficiently great to last through the season. Fresh cuttings are, therefore, sometimes seen late in the winter.

It is said that the beaver will not eat the bark of evergreen trees, and that in the Lake Superior regions no evergreens are cut, except occasionally a spruce tree which is probably cut for its gum. I have seen, and in this same beaver section, balsam, spruce and cedar cuttings which have been drawn into the water, and know that the beaver has eaten the bark from them. Upon the shores of Lake Superior the hemlock, the tamarack, and the spruce are found in great abundance. The poplar, maple, birch, and willow are also found, but not in so great numbers. Upon the tops and the bark of these trees it places its main reliance for winter food. It eats the whole of the tips of the branches, the bark, and the clear wood so long as these are tender, but does not feed upon the coarse, rough bark of the trunks or heavy branches of large trees, nor does it ever eat the clear wood, except as above stated, unless driven to it by extreme hunger. Sometimes no chips are found about a fresh cut tree, and some suppose that the beaver eats them. I have reason to believe that they are not eaten, but are washed away by high water.

Beavers are social animals. Two families frequently inhabit a single pond, but one family never occupies the same lodge. There are from eight to eleven in a family, namely; the two old ones, the two or three young ones, the two or three yearlings, and the two or three two year olds. They leave the parent lodge when they are three years old.

The beaver is not a handsome creature. Physically considered, it furnishes little interest to the careless observer; but when that man learns that it is the great animal architect upon which he is looking, his eyes at once betoken the desire he feels to know more of his habits and his works.

Building, repairing, and in fact almost all the work of the beaver is done at night. Repairs upon the dams and lodges are made quite largely of the peeled timber from the feed beds. Beavers do much of their work in small companies. They begin their year's work in the fall; in other words, their new year begins with the fall. They frequently begin their work in September; October and November, however, are the busy months with them. As trappers can make some fair estimates of the probable severity of the coming winter by the condition of the fur of animals, so they can of the duration of the cold season by the time at which the beavers begin their cuttings. The amount of repairs which they make upon their lodges is also indicative of the severity of the season. They use many precautions in their arrangements for their winter quartering within the lodge and in the pond, and so true to nature are their instincts, if, indeed, they may be said to possess no higher power than that of instinct, that they are rarely starved or frozen in.

THE BEAVER AS A WOOD-CHOPPER.

When cutting a tree the beaver sits upon its hind legs: its tail lies with the under side flat upon the ground, and serves as a support: the fore paws are put upon the trunk of the tree close to where it is chopping. The teeth, which have already been described somewhat, make at every blow deep incisions. The animal walks around the tree as it cuts, seldom making the incisions all from one side. When the tree is cut so deep as to cause it to lean, the animal becomes cautious, and when it falls retreats to the water, seeming to expect that the crash will bring some enemy to the spot. Two frequently work together at cutting, and occasionally some of the young ones help.

After a tree has been felled they begin to strip the tree of its branches, then the limbs from the branches. The larger the trunks, the shorter are the pieces into which they are cut. They rarely cut trees having a diameter of more than eighteen inches; six to eight inches is the average diameter. When a tree has been felled and the main branches are partially cut, they use great ingenuity in turning them. If a log is too heavy for them to turn they abandon it. When logs which might possibly be turned, but which would yet be very difficult to transport, are cut they do not roll or push them along as some suppose, but leave them. I have seen a great many of these logs lying as the beaver first left them.

They carry their sticks from the places where they are cut to the water, by putting one end in the mouth; the paws also help to hold it; the other end drags upon the ground. Having reached the water, they carry the stick in the same manner as when upon the land, and not crosswise and under the throat and pushed along as they swim as is often stated to be the case. Trunk cuttings average three and a half feet in length; pole cuttings are from six to twelve feet long. The accompanying cut represents some beaver chippings of the natural size. They show that the animal does a part of its cutting by splitting, and not entirely by chopping. They do not split long sticks of wood; but when they have cut into the wood as far as they wish to, they split it from the trunk. This splitting is shown in the figure by the last cutting or the upper margin of the chips.

They store their wood in piles in the water, a short distance from the lodge. The piles contain from one-half of a cord to five cords of brush and timber. It is said by some that poplar sticks, and other light wood they sink, by putting mud and stones upon them. I have never seen anything packed upon them, and yet none of these sticks are left to float. Let a man pick up from the bottom of the stream one of these fresh-cut sticks and he will every time fail to make it sink again. Some one should be able to tell the puzzled trapper how the beaver sinks his light timber. Piles of brush are also sunk in the water.

BEAVER DAMS.

The dam ranks first in order and in importance among the beaver's works; and its structure is more generally understood than that of his lodges, canals or slides.

The beaver is illy provided with homes ready made, either upon the land or in the water. Nature, however, does not leave her children without homes, or ability to provide such for themselves, and when she does not furnish the one she does the other. Few animals are so dependent upon themselves for places of shelter or enjoyment, and none are better able to meet the necessities of the case. The beaver is a burrowing animal, but as its natural home is in the water, there must be some water entrance to it. It is to provide, and to furnish an artificial pond as a place of safety from assault, and also a pleasanter water home that the beaver builds the dam.

Long continued, patient effort is required to erect these fortifications. Having once been built, they are maintained by a curious system of repairs. Many of them are supposed to be hundreds of years old, and some have doubtless been built thousands of years. The extent of beaver meadows, the great size of many of the dams, the growth of heavy timber upon some of these, and the hummocks formed upon them by the decay of vegetable matter, are some of the evidences of the very great age of this species of his work.

There are two kinds of dams; the stick dam and the solid bank dam. The first of these is found upon brooks or small clear rivers. The second species is found upon deeper water, very frequently further down the stream, upon which they have built a stick dam. Beavers tenant the small lakes of mountainous countries, and build dams upon their outlets; but they are found in greater numbers in the brooks and small rivers, probably to avail themselves of the current for the transportation of building and food materials. As they prefer shallow streams to deep water, the stick dam is found in greater numbers than the solid bank dam. As water will percolate somewhat through the stick work of the first species, further down the stream is very frequently found quite a large body of water held by the solid walls of the second species of dams.

Stick dams are built straight across the smallest streams. The sticks are placed upon the sides and not with one end driven into the ground. As they prefer those streams which have a hard and if possible a stony bottom, it is probable that it is no part of their plan to drive the stakes into the ground. Drift-wood, birch, poplars, and green willows, irregularly intermixed with mud, vegetable fibre and stones, are the chosen building materials. Their main object seems to be to carry out the work with a regular sweep, and to make the whole of equal strength. They lay the sticks crosswise and nearly horizontal and they keep them in their places by heaping mud and stones upon them. The mud and the stones they get from the banks or river bottom, and they carry them by the aid of the fore paws under the throat; the sticks they carry in their teeth.

Some dams are built of very small sticks, varying from one to two inches in diameter and from one to three feet in length. Sometimes they use logs, one or one and one-half feet in diameter. Six or seven inches is the average diameter of the sticks which they use, and they are generally about three feet long. The dams vary in length from one to five hundred yards. They are generally ten or twelve feet in width at the bottom, and two or three feet wide near the top; the crest of the dam is, however, very narrow. The height of the dam varies according to the depth of the water, varying from one to six feet; the average is three feet. The central portion of the dam is called the main dam; the extremities are called the wings.

The water must be sufficiently deep to prevent freezing at the bottom. On the lower face of the dam the sticks are arranged promiscuously, and apparently with such looseness that it sometimes seems as if the dam could be very easily picked to pieces. The sticks, poles, and logs are, however, almost always laid with their lower or larger ends against the ground, and their upper ends are elevated and pointed up stream. This is not the way men build dams; for they put the butt end of the log up stream, and the upper end with the branches down stream. The beavers are doubtless the better builders; for when water overflows its dams it does not flow into, and thus wash away the coarse cement which holds together the lighter parts of the foundation. The water face is found to present a smooth earth wall; for the interspaces on the water side are filled in with brushwood, stones, grasses, and mud. These are so well packed together that the dam resists the pressure of the water, until it rises to the summit of the dam. On the water face of the dam, the poles and sticks do not come quite to the surface; the crest is formed of mud, fine sod or vegetable fibre. Through this crest the surplus water is discharged. The discharge is regulated by the beavers; else the wear of the upper face, by the running of water through the mud and other soft stuff, would bring the water level too low, or by an overflow above the crest would wear the dam needlessly. Ordinarily the water level of the pond is uniform. Sometimes water will overflow a dam at a depth of several inches.

Many dams are so evenly constructed that the crest presents a thread-like appearance. Sometimes the crest is as even, but is wide enough to furnish a sure footing all along its line. Beaver's tracks are often seen upon the crest. Sometimes the upper margin is quite uneven, and is here and there inundated; yet this is not often the case with stick-dams in good repair. Some dams will bear the weight of one man, sometimes of several men upon the crests of the main dam, but not often at their extremities, where they are either more likely to be neglected, or where there has not yet been time for enough repairs to make them very firm. The crest line is generally at the water level, or, in other words, is so evenly made upon the surface that the water, should there come a freshet, will overflow the dam at an even depth across its entire length. Possibly beavers know that the deepest water produces the strongest current, and avoid unnecessary breakage by building an even crest.

The mode of structure varies with the character of the stream. A great number of the dams are not built straight across their streams, but are made with a curve in some section of it. The curve is almost always found where the channel is deepest, and the current the most rapid. The convex side is generally presented to the current. This is a curious feature of the curve. It is doubtless the work of the beaver, whether designedly or not; for the dam is as solid and as impenetrable at the curve as at any other section of the main part of it. If it were the work of the water alone, it would probably be sinuous in its character. The fact that the convexity is generally up stream, gives weight to the opinion that the beavers intended the curve to serve as a fortification against the increased force of the water. Sometimes the convexity of the curve is down stream, and in such cases the violent flow of the water may have mastered the beaver's first intention.

The stick dam requires the constant oversight of its builders. The repairs are so well kept up that it seems as though it must be the case that any breach, or any weak spot in the structure must be attended to by the first beaver that notices it. Sometimes after a freshet, or after any serious accident has befallen the dam, several of them may be seen working together to restore the breach. They are as watchful lest harm should come to their fortifications as they are assiduous in effecting thorough repairs when once those walls are broken.

THE SOLID BANK DAM.

The solid bank dam is in all cases an old one, being originally a stick dam, but grown solid by the gradual but continuous decay of vegetable substance, and the packing of brushwood and mud. It is generally larger than the stick dam, because of its numerous repairs. At a distance it appears from its lower face to be a very good stick dam, for so long as it is not deserted, fresh cut logs and sticks may be seen upon it. Sometimes, though not generally, the finer stuff is used in so great abundance as to make the lower face more nearly like the upper, and to present quite a smooth surface, with only here and there a log or pole protruding. The crests are generally wider than on stick dams.

When a stick dam has become very perfect, and the body of water in the pond quite deep, as repairs are still made upon it, a greater proportion of brush and small stuff and of earth are used. These being well intermixed form a coarse cement; as more of these stuffs are added, the older cement is packed harder and harder, and by the continuation of such a process there comes to be the solid bank dam. A dam which I have often seen in the Lake Superior region has a growth of cedars upon it, which will average a foot in diameter; water falls over it at a height of six feet. This dam was made at the head of a creek, and all of its water came from one small spring. Many dams have a heavy growth of timber across the entire length of the dam, and the trees and undergrowth is so dense upon the crest that it is only by cutting away the brushwood that traps can be set upon them.

Water cannot penetrate the solid bank of this species of dam; the surplus water must therefore be allowed some other passage. To do this, a section of the dam, sometimes several feet in width, is so built as to be several inches lower than the main dam; this lower section is built against the main channel. Upon the lower face, outside of the solid wall, sticks and poles are placed in the open stick-work fashion, to prevent the rapid wear or breakage of the dam. In ordinary water the current divides a little above the dam, and flows over portions of the wings of the dam. In very high water, dams are frequently overflowed across the entire length, but are very seldom injured by it, as the provision for the discharge of the surplus water is very good, and always with reference to the possibilities of the stream. The solidity of the embankment and the structure of the sluices are the only features of the solid bank dam which distinguish it from the more common stick dam.

A series of dams is often found upon a single stream. In the Chippewa country, in Wisconsin, at the head of Hay Creek I have seen nine dams; the first and the last are about three-quarters of a mile apart. The first two or three, or lower dams, were summer dams. They all set water back to the preceding dam. The water fell about a foot over the first, over the ninth eight feet; this one was at the head of the stream.

Beavers generally build their dams upon such streams as rise back in the woods, or near to some timber, or streams that are near to other bodies of water in some timbered locality. They have an uncontrollable propensity to dam up running water, and they may often be seen damming up very small creeks, if by that means they can get nearer to some standing timber.

In the summer they build a great many small and rude dams. The ponds which these make do not furnish them good water homes, but pleasant resorts, and good places for their summer and early fall cuttings. These dams are like the rude beginnings of the common stick dam and are called by trappers summer dams. There are no houses in them; but there are holes in the banks where they lie. Sometimes they lie on the banks above them.

Some have supposed that beavers fell trees across streams, to build their dams against them. There are some such, which are called fallen-tree dams; but the building at such places is doubtless accidental; for they do not lay poles and sticks across the streams, but parallel with the current, and even in cases where trees lie across the stream, the poles are placed in the usual manner.

BEAVER LODGES.

As the structure of their fore paws would indicate, they are burrowing animals. Their houses are found upon and in the banks of lakes and rivers, and upon islands. They are all constructed upon the same principle and are called burrows or lodges. They are called, from their locations, island lodges, lake lodges, and bank lodges or burrows. There is no difference internally between a lodge and a burrow; the terms are used interchangeably. All lodges are burrows; but not all burrows are lodges. The burrow is the primitive home, the lodge the improved home. In the latter they reside; into the former they run for safety, or at will for a short time. They do not generally stay in the burrow very much of the time. Burrows are always found where lodges are; but they are also frequently found in the banks of temporary ponds. They sometimes make shallow holes upon these banks, and put grasses and other light stuff in them and lie down upon these soft beds. They seem to like such spots.

Beavers are not as readily captured when in a burrow as when in a lodge; because the burrow itself is not as easily found, there being no superficial evidence of its presence. The entrances to burrows and lodges are in all cases on the under side and come from the bed of the stream or pond. They use great skill in the making of them. They range from three to eight in number. There is rarely but one apartment to a lodge, and there is never more than one family in a single apartment. The floor of the chamber is always above high water mark, and is raised from six inches to two feet above the entrances; it is smooth and hard, so hard that in cases where lodges have been opened for inspection, it is found not to yield to the weight of a man.

The beds, which are made of grass and light brushwood, are arranged around the center of the apartment, close to the wall. The chamber is circular in form; the walls are hard and quite smooth. They are made of a cement similar to that composing the water face of a solid bank dam. The average size of the interior of a lodge is seven feet in diameter and three feet in height. The walls vary in thickness according to the age of the structure, increasing as they do with every annual repair. Externally the lodges vary in diameter from fifteen to twenty feet. The chamber has no ventilation except what it gets through the interstices of its coarse roofing. Sometimes when the ground is covered with snow, there will be upon the lodge what the trapper calls a chimney hole. This is simply a spot in the snow which has become melted by the warm breath of the animal finding its way through the roof.

The external appearance of a lodge seems to the casual observer like a mere pile of sticks and dirt; its symmetrical dome-like shape, however, attracts the attention, and a closer examination reveals the fact that the sticks are so fastened together, and so surely held by the the mud, as to be very difficult to extricate. It is also a hard matter to break through the roof of a lodge. Every year just before winter sets in, they lay on fresh mud and sticks. This mud is soon frozen extremely hard, and makes of the roof an impenetrable shield, through which the Indian's arrow cannot pierce, nor the glutton thrust his claws. These houses are generally from ten to twenty rods above the dam. A favorite place for building them is upon a bend in the stream, probably because beavers prefer to live near to the deepest water. They seldom go out of their lodges during the winter, except to get their food, all of which is under cover of the water, and near their houses.

Bank lodges are sometimes built close to the water, sometimes several rods from it; the only difference between the two is that between the entrances. The shape and direction of these is regulated by the relative position of the lodge and the water. Beavers build their lodges upon islands when they can do so; because their isolation gives them increased protection.

Burrows are frequently excavated under the roots of a tree, under a fallen tree, or under some rock. There are generally more burrows than lodges upon a pond, and they are numerous along the banks of their canals. I have known beavers, when driven from a pond and from their houses during the winter, to build them rude lodges in some snow bank.

CANALS.

The excavation of the canal by the beaver is both curious and remarkable. The plain object of it is to reach timber and to furnish a channel for its transportation. As has been already observed, the beaver is an awkward traveler upon the land, but is an expert swimmer. If any amount of timber that it wishes to use grows back from the shores of its pond, it would be a difficult task for it to reach the timber, and doubly difficult to drag it along upon the ground to the pond. The canal, therefore, furnishes both a pleasant route for travel and an easy method of labor.

The observing trapper of a beaver country finds many evidences of the artificial character, and the beaver-like work of the canal. The following are some of these evidences: They are not fed by springs, as they have no where upon their entire length any current; they are too much unlike other bodies of water in the marshes and low lands through which they are found; their channels are too evenly cut to be the work of nature; along the banks of newly cut canals, piles of fresh dirt are found; it is common to find along the banks or walls of the canals and the bottoms of them, the square cut ends of roots of trees, evidently the work of the beaver; the canals often terminate abruptly on high ground, where there is not the first evidence of any spring or its origin; the course of the canals evinces as much a practical design as that of any ditch made by man.

In structure they resemble ditches much more than canals; but their object is like that of the latter, that of cheap and easy transportation of heavy articles.

They are usually cut through swampy grounds, and are fed by the waters of the surrounding marsh, by filtration, or when cut through dry land, and from stream to stream, they are filled by the waters of those streams. They are frequently cut across the bend of a stream, evidently to shorten the distance of travel when carrying logs for building purposes. It is common to find dams built across the larger canals. This is done when the canal is run upon high ground, the dam holding the water at any considerable change in level, prevents the draining of the channel.

They are about three feet wide, and three or four feet deep, and carry water varying in depth from twelve to twenty-four inches. Canals are frequently branched. In the fall of 1873 I saw a canal on the west branch of the Esconauba river, which was twenty-five rods long. It had doubtless just been completed: fresh dirt lay on both sides of the canal. The average depth was eighteen inches. Their object was to get back to some willows. They had four cords already sunk in the pond.

It seems wonderful that any member of the brute creation should have planned and executed such work as did the beaver when it built the dam. Why not live in natural ponds, or in larger streams instead of going to the head waters of small rivers, and there be obliged by such difficult means to house and feed itself? If animals have nothing to do but to eat, drink, and be merry, why should this one move back into the woods, and like his neighbor, the pioneer, be obliged to chop and dig, chop and dig, for every foot of ground he gains, for every comfort he desires? The beaver is no lazy lounger of aristocratic circles, no foolish slave to labor, but a happy worker, and the master of its field of labor; because it executes and also plans the work. It would be unjust to say that men who rack their brains to find out some new invention by which to lighten labor, that those who destroy plan after plan in order to find one after which to build for themselves the most convenient, and in every respect the most highly approved homes, are haters of labor. It is one thing to accept one's condition with cheerfulness, but it is another and a better thing for one to use willingly and effectively all of his God-given faculties for the improvement of his condition. We expect men to use their reasoning powers, and blame them if they do not; but we do not expect animals to exhibit the power of true intelligence, and when we find one that comes so near to it, and how near it is, who shall say? we look with wonder. The erection of the dam itself imprints within our hearts surprise and admiration, but its other works underscore that surprise, and make our admiration doubly emphatic.

MEADOWS.

Beaver meadows are swamps which are adjacent to their ponds, and are the result of the rise and gradual overflow of these ponds. As dams are continually repaired, and consequently enlarged, the depth of the pond is proportionally increased. If the stream upon which the pond was built lay upon low land, when the pond has become deep the land is inundated far out beyond the original bank of the stream; the vegetation is destroyed, and in course of time the large trees, becoming decayed, fall: thus there comes to be a very fertile soil for the growth of a swamp vegetation. When land once dry has become thus transformed, it receives the name just given.

BEAVER SLIDES.

What is known as a "beaver slide" is a beaten path on an inclined plane, found upon the banks of their ponds or streams. If beavers can find a natural slide of land, at a point close to deep water, they use it, otherwise they construct one. Some of the streams which they inhabit have vertical banks, several feet high. Upon such a stream they are obliged to excavate their slides, which in these cases serve as their only runways from the stream. Upon streams which have well-defined banks, and knolls here and there close to the water, beavers have their slides from these knolls. They are often seen lying upon the ground just above the slide, basking in the sunlight, and rolling and tumbling about as if it were great sport, and evidently enjoying themselves as much as a family of kittens at play. They also seem to enjoy plunging into the water from the slides.

Upon the banks of any body of water inhabited by beavers may be seen their trails. These in many cases become by long use well beaten. They are not numerous far from the shore; for if the timber is far off canals are constructed, and no trail is made.

There are places along beaver streams called by hunters feeding beds. They are simply those places where beavers are in the habit of eating, and where they have left the sticks, the bark of which they have eaten.

They generally feed in water two or three inches deep. The piles of sticks which they throw out sometimes get to be very large.

There are also musk beds near the water, which in their appearance more nearly resemble a miniature lodge. When a beaver has excreted any of its musk, or what is generally known as castoreum, it covers it with earth and leaves. Beavers from different lodges are attracted to the spot by the scent, and as every one covers its own excretion, the bed, or mound, comes in time to be quite large. They use their tails to pack and smooth the leaves and dirt which they put upon the heap. They delight in playing and in resting upon their musk beds.

The beaver is exceedingly shy, and though not as cunning as some animals, its amphibious nature gives it an advantage over its pursuer in almost every case. Sometimes, though rarely, a whole family may be caught within a short space of time. Its sense of smell is acute, and no animal is more afraid of the scent of man than this one. Great precaution is therefore necessary in the pursuit of this wary creature.

If any man were to walk around upon the path of a beaver, where it is at work upon a dam, it would doubtless leave its work for a time; it will be gone for a week or two, or more, and if the dam had just been commenced, they will all be very likely to leave it; and will, if they choose, build one in another place. Careless and inexperienced trappers will walk upon a slide or castor bed when setting their traps, and yet expect the beaver to come to the bait just as soon as if his foot had not touched the spot. Care should be used not to take unnecessary steps when setting a trap on the land. It would be impossible to set a trap well and the place still be uninfected by the scent of man; but when the trap has been set let the trapper leave the place for a week or ten days, and not expect to catch his game within that time. If the trapper goes daily to see if the animal is caught, he decreases by every visit the chance of catching his game. It will doubtless return after a few days, and if the scent has entirely disappeared will probably spring the trap.

Trails, musk-beds, and feed-beds are all of them good places for setting a trap. The musk-beds are the best places for fall and spring trapping, when the water is not frozen. Trappers look for the feed-beds a little above the dams, near the lodges. In setting traps very near the water or very near the land, or upon dams, it is best for the trapper to go in a boat, if he can do so; but even then he should be careful not to handle the limbs of trees or anything that will by the scent reveal to the beaver his presence. When trapping on shallow streams, hunters frequently wear long rubber boots, and wade; this plan is advisable.

The crest of a solid bank dam is one of the best places for trapping. A small breach should be made upon one of the runways, which are easily found in the crest; the trap should be concealed beneath the breach; the chain attached should be fastened in deep water. The ever watchful creature becoming aware of the loss of water from the pond will come at night to make the necessary repairs. As soon as entrapped it will plunge into the pond to escape, but being held by the chain is drowned. When disturbed by any means, beavers have a habit of running up and down a stream, like one pacing a floor in great excitement. It is well to set some traps on the banks, while others are set upon the dams; for if they have been disturbed by those on the land, they will run up and down the stream and over the dams, and thus increase the chance of capture.

The sliding pole or the spring pole is used with good effect in beaver trapping. It is not advisable to set traps at the entrances of lodges, especially if stakes are driven into the ground to oblige them to enter through a narrow pass; for while swimming the fore feet are laid back upon the body, and before the hind legs have reached the trap, the body, which is too broad to be caught by the jaws, has sprung the trap.

Traps are sometimes set where they draw in their timber. It is not a very good way, as it is likely to disturb the whole family, and when all are disturbed they will leave.

The best method for winter trapping is to cut a hole in the ice a little above the house, and to insert several fresh-cut sticks; they will be sure to come after them, as they prefer them to any of the old or possibly sour sticks of their winter supply. The end of a dry stick should be slipped through the ring of a trap chain, and driven into the bottom of the stream; the upper end should come above the surface of the ice. The trap should be dropped into the water, and placed inside of the coop of fresh-cut sticks. The dry stick will soon be frozen in at the top. The beaver will try to loosen the stick, or to cut it off at the bottom and at the top, and in this effort and the effort to carry off the other sticks it will doubtless spring the trap. The unfortunate animal will be caught by one of the hind feet, for the fore feet are engaged in pulling upon the stick. If the water is too deep for the stake to be driven down straight, a larger hole should be cut in the ice and a forked stick put in in a slanting manner, and the trap placed on the fork about a foot beneath the surface. Fourteen inches is a good depth of water for the preceding mode of trapping, a greater depth for this method. The same principle is sometimes carried out, but where the pole is fastened by being stuck into the bank. If this last method is used, it may be done the most effectively where the ground is not frozen at all or not frozen hard.

If bait is used, it should be the musk of the beaver, as there is nothing better than that to attract the animal, and it is easily obtained. I generally take the musk sac, and squeezing some of the musk out onto my paddle spread it upon the musk bed.

XIX.

SPECKLED TROUT.

BY W. P. CLARKE.

The American Brook Trout (salmon fontinalis) is found in the northern portion of the United States and in Canada, and his habitat extends from Maine to California. Given clear, cold, soft water, flowing swiftly over a gravelly or rocky bed, and the trout abounds and flourishes; but he abhors mud, and warm, stagnant, or hard water. The trout is found as far south as the mountainous regions of North Carolina, and is especially abundant in the small streams of the Rocky Mountains. Their weight as ordinarily caught varies from four ounces to a pound, while in some localities they occasionally grow to the weight of three or four pounds.

The spawning season of the trout, like the whole salmon family, is in the fall, in the months of October and November. As they will not take the bait until the water is clear from the melting snow, the season for fishing is during the months of May, June, July and August. None should be taken later than the first of September. There are various modes of catching them, but all depend upon the fact that the trout is blessed with a voracious appetite. The scientific angler equips himself with a fancy rod, a fine gut line, a patent reel, and a multitude of artificial flies, and in favorable localities uses these to good advantage. But there are many

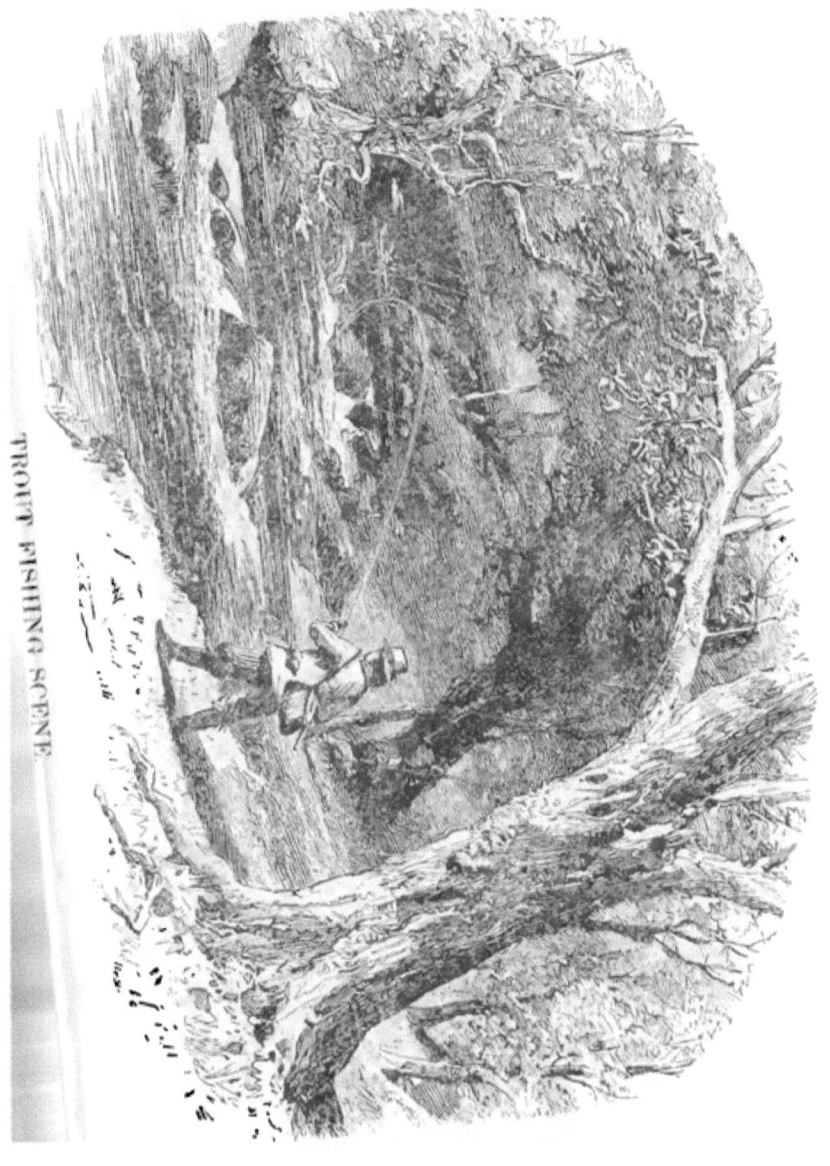

TROUT FISHING SCENE.

streams in which the speckled beauties are swarming, where this sort of fishing is impossible. In the newer portions of the country, where the trout are now the most abundant, the streams are so overgrown with trees and bushes that fly-fishing is, to say the least, very difficult. Our directions for catching this beautiful and gamey fish will be given rather to the amateur and the tourist than to the scientific sportsman; and may, perhaps, prove valuable to those who fish for profit. For fishing in the larger streams, a long, light pole of cane or other suitable material is needed, but in the small brooks where bushes overhang the water, a light sapling about eight feet long is better. A linen line, strong but small, tipped with about twelve inches of gut, called a "*snell*," a supply of No. 1 Limerick hooks, and a light fish-basket form the needed outfit.

For bait, worms may be, and are very generally used, but they are too bothersome to handle. Grasshoppers or any kind of fresh meat will answer, or the ventral fin, or gullet of the trout. But the best bait, when obtainable, is the *chub*, or the shiner. The upper part of the chub is a leaden color, and this cut into pieces three-quarters of an inch long, is greedily taken by the trout, and is easily handled. In large streams, during low water and hot weather, the trout congregate at the mouths of little brooks, and where springs make in, seeking these places for the cooler water. At such times it is useless to fish elsewhere. In a fair stage of water, however, you should not fail to try the riffles at the foot of every little rapid, as the trout often congregate in large numbers at such places. In the small brooks they will be found in the deeper holes in the bends of the stream, and under overhanging banks, logs and bushes, seeking these places for shade.

Often the best mode of fishing in these brooks is to wade down the middle of the stream, keeping the bait well in advance of you. When the water begins to rise after a low stage, the trout move up stream, and continue this until after the spawning season, when they begin to run down again. In fishing for trout the bait should never be allowed to rest, but should be constantly but gently kept in motion. In approaching the points in the stream where you expect to find fish, caution should be observed. A loud noise, breaking a dead limb, loud talking, or, if in plain sight, quick movements will often alarm the shy fellows, and cause them to dart away. Notwithstanding their shyness, however, a slight noise and agitation in the *water* will often attract them from a considerable distance. In *brook* fishing, if the trout does not bite after making two or three casts, you may as well move on, for he is either not there, or is not in the mood to take the bait. On larger streams, when fishing in deep water and in a favorable locality, it will often pay to work ten minutes in a place without a bite, as they will sometimes refuse to notice the bait until coaxed awhile.

Very good fishing is often found during the winter in the little lakes, through or into which trout streams flow, by merely cutting a hole through the ice and dropping in your hook.

To preserve the best flavor of this best of all fresh water fish, they should be dressed soon after catching. In hot weather you should stop once in two hours, or oftener, and by slitting the fish open on the belly from the throat to the vent, and, cutting loose the gullet, remove the entrails and wash the cavity thoroughly.

Should you wish to preserve them over night, or longer, and have not ice, rinse them again in cold water, spread them upon a board covered with freshly cut grass, raised from the ground so as to allow a free circulation of air. Lay them upon the grass so that they will not touch each other, cover an inch or two deep with grass, and sprinkle with water. You will find them in the morning fresh and hard, and may keep them some time longer by repeating the process.

As to good localities for this sport, they are numerous and well known. We will only mention one, which we have tested personally, and which, since the tide of summer travel has set so strongly toward Lake superior, will be found convenient to many.

The tourist, landing at Marquette or at Escanaba, will doubtless visit the iron mines at Negaunee. South from there, twenty-two miles by rail, is Smith's Mine Station. There he will find a competent guide and mentor in the person of some hunter or trapper, preferably "David" Cartwright, or "Uncle" Clemens, who will take him in charge.

I cannot, perhaps, better illustrate the opportunities for sport in that locality than by relating somewhat of my own experience.

It was a bright morning in July that "David" and "Uncle Clemens," together with my friend Paul and myself, who were stopping for a few weeks with them, left our shanty to go about three miles to a little brook to catch fish. About three-fourths of a mile above its mouth a family of beavers had built a dam. We began fishing above this dam, where the stream was about eight feet wide, and in the slack water we found glorious sport. The speckled beauties took our bait of fresh venison voraciously, and we caught in two instances over thirty from a single hole.

We began fishing at eight A. M., stopped an hour for dinner, and quit at two P. M., having caught in all one hundred and eighty-six trout, averaging about one-fourth of a pound in weight, eighty-eight of which I caught myself. Some ten days after, we visited the same stream, with nearly as good success. Three weeks spent in that healthful and exhilarating atmosphere, engaged in this exciting sport, brought renewed vigor to body and mind, and fitted us to resume the duties of life with strength and energy.

PART II.

NARRATIVES OF PERSONAL ADVENTURE.

1.

MY FIRST FOX HUNT.

"There's a trick in all trades," and there's always something profitable for one to learn. There's something for hunters to learn, not that they may be tricky, but that they may be successful. I do not know that they are proverbially tricky; but they are said to be credulous fellows, and as wild in some of their notions as are their lives in the woods, as extravagant in their story-telling as they are, when in the woods, destitute of the refinements of cultured society. There's much to be learned in the practical experience of the woodsman's life. There are the general principles of wood-craft to be understood, and there's progress to be made by way of improved methods of hunting and trapping. Besides this, just as surely as fear makes a slave of a man, to the extent of his cowardice, just so surely is a practical woodsman unsuccessful in his business in so far as his fear is the result of ignorance of the homes and habits of animals among which he travels.

My first experience in fox-hunting furnishes a good basis upon which to rest these remarks. I was but a lad, and hunted a whole day to catch a fox, but without success. The fox, true to its prevailing instinct, was too cunning for me. I came very near freezing to death that day, not as so many do, when they say of it, "I thought I should die, I was so cold," but near night I was really in a dying condition, because of the exposures of the day. I did not know it at the time. I knew but little of the symptoms attendant upon one when literally freezing to death, and I was ignorant of the best methods of self-protection, when subject to severe exposures. I was like the little boy, who said of himself after he had eaten forbidden fruit, lest it should make him so sick that he would die, "I did eat some and I didn't die any." I froze some, but I didn't die any; neither have I ceased to laugh at the fun I had, followed later in the day by such mountains of trouble as I thought were falling upon me. I have not forgotten the solemn pledge I made with myself, never to go out hunting again. The morning found me well. My determination not to hunt again had fled with the darkness of the night, and my zeal for a renewal of the chase was as bright as the light.

II.

MY FIRST BEAR HUNT.

In 1833 I went to visit an uncle of mine, who lived in Allegany county, New York. He was seventy years old, but had not yet given up his habit of hunting. He asked me one day to go out deer-hunting with him. I borrowed an old-fashioned flint lock fowling piece, and started out for some fun and some game.

My uncle told me to go up onto the hill, and he would hunt about its base, for by that means we might both of us get a shot at a deer. If he should see one he might drive it towards me, or I might drive one towards him. I left him at the foot of the hill.

It had been raining; but the rain had ceased and the wind was blowing very hard. I went on, possibly a mile and a half, watching on all sides of me, and full of expectation. I was listening too; I didn't know for what exactly. I suppose I was simply making the best use of my senses. I loved the woods. My hopes were high of some day becoming an expert in the art of catching the fish of the sea, the fowls of the air, and the wild beasts of the forest or field. My prospective field of action was broad; but I did not think it was too broad for successful work. I was fast learning, and with the exhilaration of my thoughts, I pressed my way up and on. I said I was listening. I told the truth; but I did not tell you that there was an undercurrent of tremulousness, which I suspected, and partially owned to myself, was the outgrowth of fear, and of inability to meet any bestial foe.

I heard a growling. I thought it must be a bear. I knew that a deer does not growl, but that bears do. But I said, "I'll kill it." I advanced towards the growling bear, and as I neared it I knew there must be two of them. I prepared myself for a severe engagement. I went on further. I knew that the bears were a little over the hill. I proceeded with great caution and little noise, lest I should bring old bruin from his den, and having advanced but a few steps I decided that there were three or four of them. Worse and worse. My next resolve was to get my uncle to help me. I crawled back a few rods, and hoping by this time that they would not hear me run, I went on some fifteen or twenty rods to call to him to come. But at that moment I heard the growling close behind me. I was terribly frightened. What should I do? What could I do? What could any one do under such circumstances? Three or four bears close at your heels and you standing still! I know you wouldn't stand it. I didn't either. I took to my heels. Having run very fast, I supposed I had gained so much upon my pursuers, that I could safely give one backward glance to see how they looked. My imagination was doing its work fast; it had so absorbed my attention that my eyes were almost blinded to the spectacle before me. I looked again. I listened. This time I thought the noise proceeded from a tree-top. The bears could not have been so slow in running as I supposed, and one of them must be playing sharp on me, and was just ready to light on my luckless head. Oh dear! Oh dear! What should I do? I was a dead man, I knew I was. Just then Dame Nature came to my relief and snapped the cords in my ears, so tightly strung by my imagination, and removed the film from my eyes, spread over them by the same visionary thing. This she did that I might, untrammeled by my fears, help my own self like a man. Close behind me, and in close proximity to each other, there were two lofty pine trees. There sat, like a fate watching my flying steps, the infuriate wind raging through their branches.

'Twas the first bear I had ever killed, and, from that day to this I have never been killed by a bear.

III.

HUNTING IN ALLEGANY CO., NEW YORK.

Soon after my bear scare I had another green-horn experience of a different nature. I had my first encounter with a hedge hog. I was badly frightened. I had never seen one, and when a Mr. Albright, who was near me in the woods, asked me what I shot at I told him I didn't know, unless it was the devil. Of course I did not kill it. Mr. A., if I am not mistaken, grew fat on his laughter at my performances. The hedge hog is seldom afraid of an antagonist, and why should it be, since it is so well armed for self defense? When meeting an enemy it will put its head down between its fore legs, its quills, meantime, pointing in all directions. Nothing can touch one of them without getting a sharp shower, not as when it rains pitch-forks, as some roughs would have you think, but literally a sharp, smart shower of quills. It will sometimes strike with the tail, and drive the quills deep into one's flesh. It lives on hemlock boughs, maple twigs and bark, and other wood. It is a great pest about a hunter's camp; for it will gnaw almost anything that can be found in a camp home.

Mr. A. kindly notified me that it would be better for me to keep by him, unless I was acquainted with the woods. We were only a little ways apart as we went on; soon I heard something in the brush and called to Mr. A., saying that there was a wolf close by. We turned one side to find the wolf, and found that my wolf made a deer's track. I scarcely knew one thing from another of all that creeps or roams about the woods. I saw no other game that day.

I had heard that a man could run down a deer and catch it. Some time after the day described above, when the snow was about a foot deep, I determined to test the matter. I started on my chase before it was fairly light. I soon saw a deer's track, and followed it for a long, long time, paying no attention to anything but the track. I went over hills and across valleys, and in the afternoon I came in sight of the deer. About the middle of the afternoon I heard a gun, and soon after found the deer had made a sudden turn. There was blood upon the track. A man coming up just then wanted to know if I was the wolf that had been chasing the deer so long. He asked some questions pertaining to myself, and advised me, if I was unacquainted with the woods, to go down to a certain creek and follow it until I should come to a settlement, and there take the road to my home. I had given him a sharp looking at, and had decided that I was afraid of him, and that I should take my own course. He was bare headed, and had on ill-looking clothes: he looked pretty rough, and was, withal, a little cross-eyed. I doubted his ability to give me straight directions. I afterwards learned that he was a very fine man. I went according to his directions until out of his sight, then I took the road which I thought was the right one to take. I traveled until it was dark, and was still in the woods. Having decided that I must spend the night in the woods, I attempted to build me a fire. I had put a tow and linen frock over my clothes to make me the color of the trees. Very wet snow hung on the bushes, and would fall onto me as I touched them in passing along. My clothes had, therefore, become very wet. I succeeded in finding a dry tree of which to make my fire; but I had nothing that I could use for kindling except a piece of the tow cloth, and I must set fire to it with gunpowder. But the cloth was wet; the powder was wet, and so was my gun. I could do nothing of the sort. I did not know my whereabouts. 'Twas useless to advance at such rates. Just then the clouds broke away: I took my back track and after reaching the place where I had met the hunter of whom I was afraid, I followed his directions. I found the creek and followed it. There was a good deal of fallen timber upon it; but as I dared not lose sight of its waters I walked along at a slow and tiresome rate. After a time I reached a clearing and saw a log house. I rapped at the door After I had been admitted to the room where there was a fire still burning, and had been induced to tell something of the haps and mishaps of my day in the woods, I was told to help myself to something to eat, that I should find it in such and such a place: after finishing a hearty meal, for I had had nothing to eat since sunrise, I laid myself down on the floor to sleep. In the morning I went home, and I didn't care to chase a deer again.

On my first attempt at deer shooting I was seized with what hunters call "buck fever," and I did not kill my deer. This fever, so called, is nothing more nor less than a feverish agitation which is so apt to overpower young hunters more particularly, that they cannot, even under the most favorable circumstances, control their nerves enough to manage a gun.

In the fall of 1834, having bought me a small but good rifle, I went out on the Honeoye creek with others to watch a deer lick. The Indians had just left the lick, and their fire was still burning. Our chance seemed poor, but we determined to do the best we could. I was left for a time to watch the lick, while some of the men

should go to find another one. I had been watching
about an hour when I heard something behind me, and
on turning my head I saw three deer. They were not
looking at me. I fired at them and they ran off. I supposed I had not hurt any of them. I reloaded my gun
and went up to the place where the deer were standing
when I shot at them, and found by the hair that was
lying on the ground, and by the blood upon their tracks,
that I must have hit one of them. Do you see how inde-
pendent young hunters suddenly become, and how angry
they will get when dictated? One of my comrades, seeing
me, came to me, and, finding that I had wounded a deer,
told me to go back and watch the lick, and he would go
on and find the animal; for he was used to it, and I,
poor me! was not. I told him to go and watch the lick
himself, if he wanted it done; I was abundantly able to
look after my own game, and without his help. I went
on a few rods and found my deer dead. I had shot it
through the heart. It was the first one I had ever killed,
and I was, of course, hunter-like, delighted with my success.

A few weeks after killing my first deer I was anxious
to go again to the lick to watch for them. I was then
afraid to be alone in the woods at night; but as I could
not get any one to go with me I went on alone, determined to learn to be courageous and to be a good deer
slayer. The night was very bright. Soon after the sun
had set I saw what I thought was a deer. Aiming as best
I could, I fired my gun and the creature went on up the
creek, crossed it, and went up onto the hillside, which
was there very steep. I heard it making a good deal of
noise, and in my excitement in reloading my gun I broke
the rod and it was, as it seemed to me, a long time before
I was ready to shoot at it again. On looking up the
creek again, I saw another creature like the first. I fired

at it, and with the same result as before. Supposing that
I had not struck my game, I sat down to watch for one
that I could kill. The gnats were so troublesome, that
after a little while I covered my head with a blanket
which I had with me, and lay down. I slept, but was
after a time wakened by a noise close to my head. I saw
another deer, fired at it, and, like its predecessors, it ran
up the creek. This time I thought 'twas my turn to go;
so I followed up the creek and the bank where they had
all gone. As I was climbing the hill, a deer started to
come towards me. I picked up a stone to throw at it;
for I had left my gun behind me. I do not need to tell
you that I was a greenhorn hunter, you know it already.
As the deer passed me I struck it, and it fell into the
water. I had already used my last ball. I caught the
animal by the head and held it under the water as long as
it kicked; then I drew it out upon the land. I had no
knife with me, as of course I should have had. In the
morning I got one. I also found then that one of the
others was dead, and that the third one was badly
wounded. I killed it, and cutting off the saddles I took
them home.

I killed one or two more deer, then I lay out one rainy
night and took such a cold that I came near to losing my
life. As soon as I was able to be out again, I one day
shot three ducks at one shot. On going into the water
to get my game, I again took cold and was sick for a long
time in consequence of it. The winter found me in debt.
I had never earned anything by hunting, and yet I did
not really feel able to hunt for pleasure. As I was one
day turned off from my work by a snow-storm, I took
my dog and went out to catch a fox. I soon found a fox
track, and when I had started up the little fellow I set
the dog upon it. He ran about a mile and caught it. I

could find a good market for fox skins, and I therefore continued through the winter hunting them. I got a dollar and a half or a dollar and seventy-five cents a piece for the skins. I caught two or three almost every day, and after a time I paid up my indebtedness. I owed no man a dollar. I had found what I took to be my calling, and determined to hunt for a living, whenever I could do it to advantage. I had also caught during the winter a good many coons, muskrats and mink.

Many claim that wolves will not touch a fox. This is a mistake; they will. I was one day following a fox, and three wolves came onto the track. I followed the four about a mile when one wolf turned off to the right of the track; another turned to the left, and the third kept upon the track. They ran in this way sixty or seventy rods, and the two outside ones came back onto the track, and closing together they fell upon the fox and tore it to pieces. There was nothing left of it, but a little of the fur. Another time I was following a fox and a wolf came upon the fox. The latter went into a hollow basswood tree. It was snowing, and I knew that the wolf had not been gone away long. I cut down the tree and got the fox. I once found a fox in a hollow log, and know that a wolf had been there and had tried to get it.

My dog was a cross between a newfoundland, a bloodhound, and a greyhound. While I had him, I never lost a wounded deer. I have also taken him a great many times to find wounded animals for others when they could not find them.

A young man from Massachusetts who had never seen a deer was one time visiting me and wanted me to take him into the woods with me that he might see one. I said to him after we had started out, that if he would go up the hill to which we had come, and would get into a sugar-trough which he would find at a certain place, for I did not dare to trust him to stand on the ground, for the fallen leaves made the woods "noisy," as the hunters say, I thought I could scare one up to go past him. I told him when to shoot it. As I reached a place where I thought I should find them, I saw five deer running just as I had hoped they would. Well, I thought, he will surely see a deer this day. I fired my gun, and called to him to be on the watch. I reloaded my gun and went on, hoping to get another shot at them. Hearing nothing from him, I thought possibly he had not seen them, and on nearing him I saw, — well, — the old lady had it about right when she said, "We all have our queers." Some of those queers we overcome by experience, and some of them are never bettered by this hard master; and, again, some of our queers that come of inexperience are strangely alike in many people. I saw him, hat in one hand, and gun in the other, flying in hot haste after the fleet creatures. I asked why was all this? He saw them and when they were within a few rods of him they turned about to see what had frightened them. He thought he could catch them, and forgetting that he had a gun, he started with it, as I have described, after them. He did see some deer and thought them the handsomest creatures he had ever seen.

After that, another young man who had never seen a deer, came to see me. My directions to him were quite similar to those given to the one just referred to. Very soon I heard his gun, and on going to him I found him with his coat off, and working for dear life to reload his gun. He said there were two lying near him, and that he had shot one of them through the middle of the body. He offered to get the dog, for we would need him, if I would load the gun; he was so excited that he could not do it. I found the deer about twelve rods from where he had shot it. It was dead; he had shot it through the heart. When he came back with the dog and learned that he had killed the deer, he jumped about like a crazy fellow, and, shouting at the top of his voice, gave utterance to some not so reverential cries of gratitude which we will not here repeat. He afterwards went with me on a fox hunt and then gave my dog, whose sagacity he had disparaged, all the credit I could ask one to give to such a faithful and efficient huntsman's dog. In a deer hunt, which I had soon after that fox hunt, by an accidental shot I crippled my dog; but it lived and did me good service many a day.

The first time that I left my home to hunt, I went into Pennsylvania. There were several men with me. We went into a rough country, about seventy-five miles from home; but it was a good country for a hunter and trapper. We hired a man to carry us as far into the woods as he could conveniently. About as soon as we were well fixed for camping in the woods, some of the men became homesick and wanted to turn about at once. There were bears, wolves, panthers, martens, otters, mink, the Virginian deer and the elk, and other game in abundance, and it was very foolish for any men to leave such a place, if they wished to become practically successful as woodsmen — more than foolish; it was, as it will always be under similar circumstances, impossible for them to succeed at such rates. Hunters should be as diligent in business as any other class of men, if they wish to succeed, and it is my opinion that more discredit is brought upon woodcraft by a want of judicious persistence in the business than in any other way. I came into business relations with a man, during the last season of my hunting in Allegany county, who very soon became tired of the business we were doing, and who wanted me to go on with it alone, and let him spend his time in hunting. The arrangement was made as he proposed; but, for lack of good management in his hunts, when he left it some weeks after, and I also stopped my work, it so happened that he was one deer behind me. In 1841 I did my last hunting in York State.

IV.

AN ADVENTURE WITH A WOLF.

In 1842 I moved into Jefferson county, Wisconsin, into a heavy timbered section, known as Bark Woods. There was no settlement there at the time, and no roads. I built me a house on an Indian trail. The wolves were then more numerous than I have ever known them to be elsewhere. We had to keep all of our young stock shut up, to keep them from the wolves. It was not uncommon for these fierce creatures to chase a dog to the door of the house, which, by the way, was not furnished with doors, only blankets stretched across the door-way.

The first winter that I lived there, when it was one day snowing, and the ground already had six or eight inches of snow upon it, I took a trap and some pieces of venison and went out onto the trail about sixty rods from the house. I set my trap upon the trail and returned. At midnight we heard wolves howling unusually loud, possibly the loudest I had ever heard them. I rose early and found that a wolf had carried off my trap. I went home, got a chalk line, a rope, my dog, and I got my neighbor, a Mr. Thomas, to go with me and help to get the wolf home alive. I wanted to do this because some of my neighbors, who had recently moved into our settlement, had never seen a wolf. We followed the track for a half mile, when we found the wolf lying by the roots of an old upturned tree. This tree lay by the edge of a thick tamarack swamp. The chain was broken, but the clog was hitched. Thirteen wolves had followed the one which was entrapped. When we came in sight of him I saw him go into the swamp. He had heard us. I set the dogs on him, and ran on myself as fast as possible. I had outrun Mr. Thomas. The wolf had whipped the dogs, so I was there to face the danger as it might come. It did come. The dogs turned about and ran towards me: the wolf followed up the dogs. I caught hold of a little tamarack tree, and broke it off by the roots, which were rotted: the top was dry and sound. As the wolf came in reach of me I struck him on the back of the neck. He fell over a tamarack root, and, as he fell, I put my tamarack pole over his body, fastening the further end of the pole under the root. I held my wolf there, as in a vice. Mr. Thomas took my place and held the pole down, while I took my line and tied his mouth. I then put the rope around his neck, and I tied his feet together. I put him upon my back and started homeward, a sort of soldier, with a knapsack upon my back. I felt the wolf squirming about, and I squirmed too, as I felt his cold nose on the back of my neck, even though I knew that his mouth was fastened. I carried him home, and called my friends in to see the new arrival.

It was so cold they did not come until the following day. I therefore tied him up securely and put him at the end of the house to keep him over night. In the morning I found that his legs, his ears, and his nose were frozen. I took the line off of his mouth, and carried him into the house and put him into the wood-box. He was apparently pretty near dead. I went to hurry my neighbors lest he should die. While I was gone he got warm and commenced to look about the room. This was not to be borne by the inmates of the room. Mrs. C. yelled at the top of her voice, and he curled down in the box. By this means she kept him within bounds, while the little five year old boy ran for Mr. Thomas to come and take care of the wolf. When he came he pulled him out of the box by a rope, and dragged him out of the house. As the wolf passed the door he struck his teeth into a log, (the house was made of logs,) and tearing off the bark, cut into the log so deep that the marks of it could be seen for a dozen years. When I reached my home, and my neighbors with me, we put three dogs upon him. He whipped them all, and one of them he almost killed.

This wolf was one of the Timber Wolves, the largest species known, except the Rocky Mountain White, or the Buffalo Wolf. This individual wolf was an exception to his family, in this respect,—that he was not very tame when entrapped. I killed him. He was two feet and eight inches tall at the shoulders. He weighed seventy pounds.

V.

TRAPPING IN JEFFERSON CO., WISCONSIN.

My home in southern Wisconsin was in the town of Sullivan, or what is known as Bark Woods. There were but three or four families there when I settled in the town. The country was, however, thickly settled with deer and wolves, and with Indians. Bees thronged in multitudes of swarms, and their honey was very abundant. I commenced with my neighbor Mr. Thomas to hunt bees, and we were very successful.

One day as we were going through a tamarack swamp I noticed that something had been gnawing the trunk of a large tree. On close examination I found the marks of a bear. It had pulled out from the hollow tree a dry knot, and, looking through the knot-hole, I knew that the bear had found some honey. In the morning we cut the tree down, and took out one hundred and sixty pounds of excellent honey.

Another day we had been a couple of miles from home to get potatoes. It was very dark when we returned to our homes. When there was but a hundred rods to go before we should reach our rude but comfortable dwellings, I heard something traveling in the path in front of us. Mr. Thomas thought it was a deer. I thought it did not walk like one, but suspected the wolves were on the path. Remembering that a handspike was standing against a tree as we passed by this spot in the morning, I caught hold of it. Mr. Thomas, also getting something for self-defense, declared that he was ready for them, that he would saddle one with a bag of potatoes. When we walked they did; when we stopped they also stopped. A fallen tree lay across our path, and we were obliged to go around it. Soon after we had passed it I heard a wolf turn out of the path, then another, and another, and the fourth one. They then set up such a howling as I have seldom heard. It seemed as if the ground shook from the violence of these creatures. The dogs began to bark. We called them and tried to set them upon the wolves, but could not. On reaching my home I learned that the wolves had been close to the house, and had frightened the dogs so much that they could not possibly be controlled.

I have known dogs to be so badly frightened by wolves as never to recover from it. One of my neighbors had a dog so badly frightened by a wolf that after it found that it could not get into the house, it started for the road, and ran six miles from home, yelping at every bound as it left the house. Soon after it had started, another dog hearing the noise came to the rescue. The wolf left the first dog, and put upon the second one. This dog succeeded in reaching its home, though not without some wounds from the wolf. The other dog was found in a couple of weeks and carried home. Towards night it began to grow uneasy, and acted as if it was afraid of something. This fear increased with the advancing darkness of the night, as it came on, and soon after sunset, when a limb fell from a tree close by the house, the dog began to howl, ran out of the house, and they were unable to catch it. They could never keep the dog at home afterwards.

As soon as the weather became so cold that we could not hunt bees, we commenced to hunt deer, and we were again successful in our hunts. The deer used to go into the oak openings at night to get acorns to eat. In the morning they would go onto the swamps and stay during the day. We used to go out very early in the morning, and watch upon their runways. One time I killed four before sunrise. Another morning when we were watching, and were about forty rods from each other, just at daybreak I heard a gun. I watched closely to learn, if possible, the cause of it. I thought I could hear something, and went over the knoll that lay between us. The snow had fallen about nine inches the night before; but it was beaten down for several rods around him. He was sitting on a log and breathing hard, as though he was recovering from a fight with some animal. A deer had come from behind an upturned root, and he did not see it until it was close by him. He had shot the animal on the back of the neck; it had fallen, but seeing that it was getting up, Mr. Thomas caught it by the horns: it jumped, and dragged him off about eight feet. Just as he thought he was getting some advantage of it, the deer made another lunge upon him, and he fell as before. The deer probably meant no harm, but was trying to save its life. Mr. Thomas was also trying to save his life, by preventing the deer the free use of itself. He had, therefore, held onto the horns as long as possible. After a time he succeeded in getting the upper hand of it, and when I reached the spot the deer was nearly dead, and Mr. Thomas was exhausted. We used to leave our homes at five in the morning, and return about ten, of the same morning. We caught that winter upwards of seventy-five deer, eight or nine wolves, several wild cats, and a great many coons.

We had a deer hanging up in the woods: a wild cat came to eat it. We therefore set a trap for the cat. One morning when we went out we were obliged to carry a lantern, and, as we were to pass the trap on our way, we thought we would see if there was anything in it. I saw that something had been there. There was a large pile of brush close by, and on looking into it to see what I could find, a huge wild cat jumped out at me so quick that I jumped too. On jumping back, I hit my heel against a bush, and I fell on my back. My next motion, which was to regain my footing, was about as quick a one. The cat had carried the trap off from where we had set it, and, as it sprang for me, the chain became entangled in the brushwood, and the cat could not quite reach me. I lit my lantern, which had, meantime, gone out, and shot the cat, doing it with a vengeance, to pay it for scaring me so outrageously.

The next year we hunted a considerable, and selling my honey and my deer skins in New York State, I could replace the money that I had lost, for which I was to pay for my place. In the fall of the year following my hunting with Mr. Thomas, partially sketched for you, we went together again. We had taken a boat, and were going down on Bark river and the Scoopernong. I saw something crawling in the grass, and shot it. It proved to be an otter. I had never yet seen one, and had to be told what it was. The day following we caught three of them, and in two weeks we had caught sixteen. We also caught a good many muskrats, coons and mink.

Later in the season I shot a deer and broke its leg. It fell down, but as I snapped my gun it jumped up and started to run. I set my dog upon it, but it whipped the dog, and I ran to them, caught hold of the deer, and it threw me over a large tree that was lying on the ground near us. I arose in so great surprise that my dog caught some of my fire, and pitched into the deer again. We two excited ones killed it in a short time. On my return home I saw a fallen tree: it had fallen while the leaves were on, and as it lay across my path I could just see the tip of the deer's ear. I whistled, to get it to lift its head, and it did, but not high enough. I tried it again, but the deer did not move. I shot at it, and away went five or six deer. I went on towards home, but being anxious to know if I had struck the deer, I returned, and found that I had split its head open, and it was of course dead.

Late in February, while hunting, I came upon a bear's track, and after following it for a short distance I reached a spot of ground from which the snow had disappeared. I looked about every log, to see if I could find some track leading off from it. I could find no other tracks, so I knew the bear must be near me, hidden in some log or hollow tree. I found his tracks upon a tree, and a large hole upon one side of it. On the opposite side there was a crack in the tree nearly an inch wide. Through this crack I could see the bear's hair. How to kill it was the question. I must shoot it while it was in the tree; but when I would go away from the tree to take my aim, I could not see the hair. I returned to the tree and marking the spot the best I could without disturbing his majesty, I stepped off to shoot. I became quite excited, but succeeded in hitting my target. The excitement within was just then very great, as the growling and pounding fully attested. I loaded my gun as quick as possible, and went around to the opposite side of the tree to shoot, if the bear should attempt to come out of the hole. He did not, and in a short time I cut the tree down and found my bear dead.

One winter when a couple of men were stopping for a time near my home, to hunt, and having been two weeks without catching any game, they wanted to buy some of me. I was at the time hunting on a contract, and could not sell them any; but I told them that I would kill a deer for them the next day, for two dollars and a half. But they wanted two and I promised them the two. I purposed to go out alone; the men, however, were very anxious to go out with me, and having promised to be good boys and not disturb me nor frighten the game, I consented to let them go. One of the men soon lost us, and we saw no more of him through the day. After a long time when I saw a deer, and was about to fire at it, the man who had kept with me, got a little too close to me; he caught my arm and told me not to shoot at that distance. Before I could again take aim at the deer, it was far beyond shooting distance. It crossed a swamp, went over a ridge, and upon a flat beyond it. I told the man who was with me to stand quietly behind a certain tree until I called him. The woods in which we were, were what Western people know as "openings." I crawled upon the ground as stealthily as I could, and for some time after I had reached the top of the ridge, or bluff, I kept in range of a tree that hid me from the deer. There were several standing close together, watching a man beyond them, who was driving his cattle to water. I shot, and killed the leading deer. The others, though frightened, very soon came to look at the fallen deer, and when I was about ready to fire at another one, I looked up and saw them running away from me. I looked to see what was the occasion of the fright and saw my huntsman running towards them. I called to know what that meant, and learned to my delight, satisfaction, or what you please to call it, that he was anxious to know if I had seen that deer fall. I knew I had not seen but one fall, and yet I had expected to catch several of them.

If you want to know how I felt for a moment, I think that Gough's story about the good baby would explain as well as I could tell. When a mother was speaking in very high praise of her baby, because it never cried, nor made any trouble, some one asked her if she thought it was a bright child. I told the man he was too big a fool to be in the woods, that I should go home and leave him to get his second deer.

Independently of such poor company as I had with me on the day just referred to, I have made some very poor days at hunting; but they have not all been unsuccessful. On one hunt I caught thirteen deer in three days. I have quite a number of times caught five in a single day. Once I got six in one day, and that day at my first shot I killed two large bucks and at my fourth shot I killed two fawns. Of the last fifteen tracks which I saw in that locality I killed fourteen deer. My farm work kept me busy during the most of the year. For five or six weeks during the fall and winter I used to hunt, and I would make two hundred or two hundred and fifty dollars in that time. In 1855, when I did some of my last hunting in Jefferson county, the deer were very scarce. There have, probably, not been a half dozen killed there since that time.

VI.

A TRAMP TO CALIFORNIA IN 1852.

The desires of my boyhood were to be more than realized. My dreams of life in a wild country were to be no longer dreams. In the early spring of 1852, in company with two others, I started from my home in Southern Wisconsin, to conduct a company of men across the plains to California.

For any number of persons to undertake an overland journey to California a score of years ago was to set at naught the pleadings of anxious friends, who were to be left behind, to face a long and exceedingly tedious journey, and one fraught in many cases with imminent peril. It was to travel, even in well equipped and thoroughly furnished companies, at such disadvantageous rates as would mock the fruitful imagination of even appreciative travelers, while crossing many of those same rivers, plains, mountains or mountain passes, but who view them at a rapid rate from the windows of some Pulman palace car, upon the Union Pacific Railroad. While now, children may be put in the care of some watchful conductor, or partial stranger, and be safely conveyed in the lapse of a few days to their places of destination upon the extreme western coast; while timid travelers and infirm people have but to be seated and wrap about them, as with a garment, a spirit of fearless quiet, and bidding good-night to an Atlantic home can so soon say good-morning to the Golden Gate; twenty years ago no man ever dreamed of going alone from shore to shore, and companies of stout hearted, fearless men, many of whom were in those days, figuratively speaking, crazy to reach that land flowing with gold and money, not unfrequently lost members from their ranks, who were literally crazed by the exposures and anxieties of the journey. Then, as now, not all the men of any company that would be organized could make good pilots through a trackless or an unknown land; and then, as now, what is everybody's business is nobody's business, and this saying applied to the case in hand, to be understood, that every man for a pilot would soon leave no man for a pilot, it was safer and pleasanter for a company to be subject to the leadership of an authorized guide.

Gold, gold, gold, was the excited cry all over the land. It was the great Emancipation Proclamation of the day, and California was the Canada of the slave to poverty. It was this cry of gold that impelled the majority of the men composing the company of which I was one of the guides to leave their homes, and brave the dangers of such a journey at such a time. There were thirty-three men in that company, twenty-one of whom were residents of Sulivan. These, understanding that it was in every respect the better policy to secure one or more guides, had engaged Mr. Miles Homes, Mr. John Nutter, and myself to conduct them to that promising land. We were each to receive two hundred dollars for taking the men across to California.

These thirty-three men were all of them at the time residents of Southern Wisconsin. The State was then in its infancy, and no home in it was other than a pioneer's home. To men living in such homes as these did, where every man's neighbor lived on a footing with himself, and where, though the necessities of life might be quite well supplied, the luxuries and even the comforts of older homes had scarcely been talked of as a possible addition to what they already possessed; to these men, an overland journey to the western coast possessed none of those attractive features so inviting to the west-bound traveler of to-day. They were men used to toil and to privations, but who were bent on getting that curative of so many of the ills of life, that ever precious gold. The difficulties of the way between them and their promised treasures they were bound to meet with manly courage. What they did, and how they fared, it falls to me after the lapse of more than a score of years to tell you.

Having once decided to go, and the necessary arrangements having been freely discussed and in due time carried to completion, the company was, according to arrangement, to meet at the house of your narrator. The eleventh of March was the inauguration day of the great event. I say the great event, and I think I am justified in calling it such, to some of us, at least, and as there's no one of the company here to contradict me, I shall not recall the statement.

While I know that my neighbors are coming, but before they reach my door, I try to settle beyond a doubt, if I am ready to leave my family, in which there were several young boys, upon a frontier home, to care for themselves as best they can, and if I can believe that all will be well with them, should I never return. It was to be a journey of several months, at the best, of exposure and fatigue. As incivility is nowhere the type of a true gentleman; but as, on the contrary, a hearty welcome is the one best thing that can be offered a guest, let me put you in a position to feel something of that welcome, which I felt, as I opened the door, for some of the men with whom I went to come in.

Mr. Holmes, who was my partner in the undertaking, was born in Waterbury, Massachusetts, of excellent parentage. His family connections are well identified with important manufacturing interests of the State, and many of them are exceedingly wealthy. He was a man of strong executive ability and business-like habits. He had been for a number of years a merchant in Georgia, prior to his removal to Wisconsin, where I first became acquainted with him. He was appointed Colonel of State militia of Wisconsin, by Governor Dodge. You would like him, and were you to enter into any business relations with him, you might rest assured he would not fail you, nor would he in any way do violence to his promises; he would also, doubtless, be on the ground before you were, at the time appointed for any convocation, or for work in any shape. One must be up and dressed to get in ahead of such a man. He was stout and brave, and yet sympathetic and easily moved to tears. Apparently fearless of danger, and regardless of suffering, he was in truth as much excited in sympathy for the sufferings of those around him, as he was scornful of any coward's cry of lions in the way.

Mr. Nutter, who was the third partner, and who was elected to the captaincy of the company, may be known by his fine athletic frame and robust form, his healthy countenance and complexion, darkened somewhat by exposure to the sun, hands, and muscles generally, hardened by actual contact with the toils of life, and his dark, piercing eyes, which if they were darkened by other cause than nature's choice, must have been by the charred remains of the many fires that have burned there. He is one of your lucky fellows, and he makes those lucky, too, with whom he is associated. We shall need him many times before we reach our journey's end; and as well might one expect a Yankee question-maker to be outwitted by a California story-teller, as for a party like ours to expect to get through such a journey in safety without taking into their number this ingenious fellow, or some of his pedigree; though they may not be related to him by the ties of consanguinity, if they are by mental aptitudes, the end is accomplished; for the never failing luck of coming right side up with care, will serve them many a good turn, and disarming men of doubts and discouragements, will put in their stead confidence and good humor. If there's a river to be crossed, and no one can find a fordable spot; if there's extra baggage to be packed, and no one can find a place to stow it away; if there's a bad place in the road, and no one knows how to get around or over it; if it is so hot or so cold, so wet or so dry, that no one can tell how to adapt himself to the circumstances and keep on the march, Nutter is in his element, and no sooner is the difficulty apparent than his head and his hands are hard at work to make a way of escape. If a trail is lost and the heavens withholding help, keep back behind their clouds the twinkling stars, his head is clear, and in the twinkling of his eye you may find your assurance that all will yet be well. If you still doubt it, he has but just returned from such a trip, and knows whereof he speaks. Besides all this, he is a jolly fellow and will be excellent company upon the road.

Charles Hibbard, a man of excellent principles and upright morality, is a quiet, thorough-bred gentleman. Though by no means a conversationalist, as the world calls such, he must yet be in possession of one of the key secrets of good conversation; he always stops talking before berating his neighbors. He shows himself friendly to every one, and all are friends to him. For his cheerful, happy spirit, and consequent wholesome influence upon us, he was welcome to our number.

Charles Dunning was born of fighting stock. I think he was. Physiologists tell us of the varied forms and functions of the corpuscles of the blood; that some are red and some are white; that the former exceed the latter in numbers; that the latter are the larger, and that the red ones are doubtless the product of the white ones. Now it is certain that the blood which coursed his veins was fired by elements somewhat antagonistic. The milk of human kindness was there, the white corpuscles of his nature, a pugilistic spirit, calling for bloody revenge, if need be, the red corpuscles. The former, when observed displayed the generous, warm-hearted man; the latter, though more frequently manifested, may have been, after all, the outgrowth of the former, produced by a desire to see every man receive his just rewards, to see humanity rewarded and malice punished.

Stephen Davenport, who took with him a son, resided in Jefferson Co., Wisconsin. He had a strong inclination to corpulency; whether that inclination was wholly mental, or wholly physical, or a mixture of the two, his personal appearance and further acquaintance with him might furnish you a good chance to learn. Some one, speaking from 1874, asks if he belongs to the Fat Men's Association. We cannot tell you; but his credentials at the time to which I refer you for an introduction, would have been accepted by said Association as readily as you should now accept his company; for he was both fat and jolly, his two hundred avoirdupois being well balanced by his love of fun. His love of a horse exceeded almost any other display of his affections. Nothing seemed to suit him better than to get a good horse, and to care for it well. He was easily excited, and was then very strong.

I have known him, when under excitement, to carry weights that in a state of relaxation he could not, with the help of two able-bodied men, lift from the ground.

Of the Jaquith brothers there were three. They would go everywhere, and you would think all at once, up hill and down, upon high peaks or rocks, through ravines and across streams, to see what could be seen, and to hear what could be heard. They will make good picket men: their quick ears will catch the first sound of coming danger; their judgment, quick to act, will soon decide the case in hand. As for ability to mimic all the strange noises, earthly and unearthly sounds, there's none can beat them. Coyotes, Prairie-dogs, Buffaloes, birds upon the wing or in the woods, and Lo, the poor Indian, stand their chances alike of keeping in advance of their imitative powers. They are the fun-makers for the crowd. Should they, prompted by their native inquisitiveness, go too far beyond the camping ground, some one must look them up; else after some wearisome march, when a jolly chat over the evening meal, and about the cheerful fire, would make the men forget the fatigues of the day, and fit them for a healthful sleep, they must wrap the mantle of their gloomy moods about them, and lay them down to sleepless dreams.

Henry Torry, who was an active, nervous, plucky little fellow, was the oddest genius in the company. His drollery, of which he was sometimes conscious and sometimes not, was not to be matched by any of us. He is one of the many who have been, by some fall, knocked "sensible" in less than a minute. He says he fell from a trespass, 'twas no precipice, I assure you, and that he remained sensible for some time after his fall. He is an excellent man, and his excellence is in keeping with his zeal for the promotion of good.

If any of the company should think that I have forgotten that Abram Balsar was one of us, he is as much mistaken as if I were to say that our food never becomes a part of us physically, and that the condition of the stomach has no influence upon the conduct of the head. Balsar was by profession a baker. At the time of which I write, he was practically both cook and baker. He was also a good fiddler. You may call him a sort of homœopathist. He was, in his way. He was an expert in dealing out palatable dishes, sweet to the mouth as are sugar-coated pills; and when, after the march of the day, he had, by virtue of his office and by his willing hand, helped to place upon our rustic board what satisfied our cravings of hunger; and when there was still a hunger not touched by bread or meat, he could fill our ears with what made music in the heart, and satisfied its cravings for something restful and cheering.

If you, my readers, have also read the preface to this book, you have learned that David is not doing all this writing for himself. Therefore, seizing my chance to take a little advantage of him, I purpose here to intrust to your supposed safe keeping a secret. It concerns David. I think his motives in going, though I've never heard him say it, were his love for the woodsman's life, his chance while thus living to still work for his family, and his share in the chances of the good time coming, when they should reach the Golden State. I think, from what I know of him, that his friends wanted him to go; because his head was a capital compass; because he was not afraid of wolf or bear, or Indian, crows or scarecrows; because there was no other so good a hunter and trapper whose services they could procure, and through him procure the needed game. Should he ever learn of this, and charge the telling to me, I shall simply tell him that he owed it to himself, to tell you for himself why he was released from picket and other duties, and that, should he revise this book, he must tell you, in his own queer way, of his adventures in '52, and not be moved by that modesty that now cheats him out of saying a word in his own behalf. Just one other thing let me tell you, and we are ready for the forward march. When our hunter tells you his own story, he will doubtless call himself " David and I," as that's an old trick of his. Many of his neighbors will vouch for the truth of this, that when he leaves them he will say, " David and I must go."

As the servants of the company, there were four ox teams and one horse team. We had twelve yoke of oxen, three yoke were hitched to a wagon, and we had five horses. Four of the horses were driven, and one was used for a saddle horse. Our wagons were well loaded, and thus equipped, our merry, sober men left my house, leaving behind us a company of about two hundred, who were possibly less merry and more sober than we were. The next day two companies met us, one from Cold Springs, and the other from Lake Mills, adding to our number twelve men and nine yoke of oxen, and one horse.

The roads were in very poor condition, and the water was high ; but we went on, and across Illinois, following the eastern bank of Rock river, then on the east side of the Mississippi until we reached a point opposite Fort Madison. We would put up at night at some private house, or at some tavern, and at noon would feed our teams upon the road, and feed ourselves as circumstances made it most convenient.

Just before we crossed the Mississippi, an Irishman came into our camp, while we were eating our dinner, and without saying a word to any of us, took a plate and some victuals and seated himself to eat his dinner with us. After he had satisfied himself and was about to leave us, Dunning demanded pay for his dinner. But, says the Irishman, "Fath and be jabers, ain't I in a free country, and hain't I got a right to eat when I plase?" "Yes, and ye're in a very free country, and ye're free to pay for what ye get, too!" But poor Paddy, swearing by all that's good and great, declared that he couldn't do that, for he hadn't got any money; whereupon Dunning, either having no faith in his word, or feeling abused and determined to get satisfaction by giving Paddy no chance to digest his dinner, grabbed him by the collar, carried him to the river, and threatened to throw him in at once if he didn't fork over. Paddy, though he didn't really know that a man ever had red corpuscles in his disposition, yet appreciated the force of his determination, and handed over a five dollar gold piece.

Near the close of the day, having made arrangements with the ferry-man to take us across the river that night, we set ourselves to work to get the baggage in readiness. The boat was small and the ferry-man was obliged to make several trips before he could get us all over.

When some of the teams had been taken, and the boat was about to be reloaded, in defiance of the rule that no man or party could be ferried across until the previous engagements were fulfilled, an old Dutchman and his wife were about to drive in ahead of our men and teams, upon the boat. Some of the boys caught the horses and others caught the wagon wheels; as the Dutchman's wrath made it too hot for him in his wagon, he got out to give his wrath a good ventilation, and those impudent, ill-mannered boys a good flogging. He first clinched Nutter; but they, seeing that the struggle was unequal, lent a helping hand, doubtless with the intent to persuade him, by a practical experiment, that they rejoiced in the opportunity to rid him of such a nuisance as our Captain seemed to be to him, because he would not let the aged couple cross just when they pleased.

His wife, however, who had been left the only occupant of the wagon, looked at the matter from a different standpoint, and didn't read their intentions as I have stated them to be. She had a view of her own. I'm inclined to think she never goes anywhere without one.

If she was one of your four-handed people, as wealthy people are sometimes said to be, I cannot say, but this I know, she had a very hard fist, harder and more to be dreaded than her husband's was. Conscious of her strength, and possibly of her indignation, she jumped from the wagon, doubled her fist, and with it struck Nutter with such vehemence as I've seldom seen displayed between man and man. She was a striking character, I can assure you, not strikingly handsome, as many women seem to wish to be, nor strikingly "tame," as some of whom we've recently heard, but striking in many of her ways.

Nutter, who was now in the position of the light brigade, with

> "Cannon to right of them,
> Cannon to left of them,
> Cannon in front of them,"

for the Dutchman and his wife were all around him in no time,

> "Volleyed and thundered."

"Good heavens! woman, I won't strike you; but I *will whip* your husband." Our boys, coming again to the rescue, proved themselves the victors, and the venerable couple decided to do, as all good old people will do, patiently bide their time to cross the river of — Mississippi. Nutter did not soon hear the last of his unequal tussle with the Dutchman.

Mr. Holmes and Mr. Cole left us to go down to St. Louis, to purchase supplies for the remainder of the journey.

We went on to Des Moines, and thence across Missouri to St. Joseph. In our slow march through the State, we saw nothing, neither did we hear anything, nor did we meet, nor were we met by anything beyond the ordinary knowledge of thousands of pioneers. We were in their country, and yet in a new and sparsely settled one. We were outside of fences, but not beyond the pales of the white man's habitations, nor westward of his pale face. Houses were rude, but there were comfortable ones: there were some roads and such as we would be ashamed to underrate, knowing as we do, of the hard work that is required to make good ones in a new country. The soil, of which clay is an important constituent, was wet and sticky. The streams were high and there were no bridges. Several of these we bridged ourselves, but as we did it on foot, or on horseback, the bridges did no one any good, except ourselves. There had been as yet no pursuing Egyptians, nor red-man's host to follow upon our trail across these streams, in the vain hope of catching us on the other side, or to be engulfed in a fordable river. One we could not ford; but upon examination we found our spirits were higher than its waters; and by dint of perseverance and a little management, we succeeded in crossing and in reaching the other shore high and dry. We felled trees and run our wagons over on them, and we swam our cattle and horses across. This State, as many more inhabitants of it can testify, is in many respects physically beautiful.

Soon after reaching St. Joseph, I received a letter from Mr. Holmes, telling what day he expected to start from St. Louis, and that he would be on board the Seeloday.

Our next news from St. Louis was that the boat, while on its way up the river, had been blown up, that four hundred lives had been lost, and that all of the freight was lost. A chill of horror ran through our hearts, as it did throughout the land. Many households were to bow in sorrow, if not submission, at the loss of some loved one. Four hundred lives lost! and that our friend Holmes had escaped was scarcely to be thought of. All of the freight lost! and we knew that our food stuffs were gone. The loss of Mr. Holmes would be a doubly great loss to me; for we, as partners in the undertaking, had been bound under a forfeiture of two-hundred dollars each, to the nineteen men who started from my house with us, should we fail to take them through to California according to contract. There was not money enough in the camp with which to buy our necessary outfit, in case the worst possible news should prove to be the truth. A merchant in St. Joseph, knowing our extremity, told Nutter and myself that he would furnish us with what we wanted, that we might pay him what we could, and send him the balance of the indebtedness after reaching California. This he would do in case our men were dead, or our provisions were lost. That day a boat came in from down the river, but brought no word from Mr. Holmes, nor of him. I determined to go down the following day, to learn something about them; should it be good news or bad, we could wait no longer in suspense. A boat came up just before mine was to go down, and when it landed I saw our men upon her deck. 'Twas a joyful moment to me, and though it was a joyful thought to Mr. Holmes that he had reached his men in safety, his eyes were as if still riveted upon some awful scene. He took my hand with a warm grasp, but he was speechless. Large tears coursed unchecked over his manly face. I will not, with the sneer of some of you, call them womanish tears; for 'twas manly to weep with those who wept, to weep for those bereaved.

It was some little time before he could control his emotions, and could give us any details of the disaster. When he did, it was with no parade of sensational narrative that he recounted to us the events of the terrible catastrophe, but from an overburdened heart, still strangely horrified, that he said to us: I saw, as I approached our boat, that it was very heavily loaded. I feared that it would give us trouble; but as I was to be only a passenger, and not wishing to make myself offensively conspicuous, I kept my fears in check. After we had started, and were in a bend in the river, at which point the stream was very rapid, and the boat could not, for its freight, work its way over the waves, and after it had made two attempts and failed, I said to the captain "Throw out a line and let a hundred of us get out and pull on the rope, and help you over this." "No! I won't! I'll run it over myself, or blow her to h—l." "Have you ever blowed one up?" I asked. "Yes, I have, and I done the old thing good justice, too." I went around to the engine room, and saw that the safety valve was tied down. I said to Cole, "That boiler will burst, and we're a shipwrecked set. Let's go to the stern." We had scarcely reached it when the deck was raised, everything rose up, the boat was instantly shattered to pieces, and scattered upon the waste of waters. How we escaped I cannot tell. Hundreds of bodies of dead and dying were mingled in that sea of blood, for the blood of mangled bodies was flowing fast and free. The living were making loud moans, calling in their anguish for help; some calling for their friends, children for their parents, and the parent for her child. A mangled part of the captain's body was found upon a shed about four hundred feet from the water. He had lost everything, and there was nothing but the few shreds of clothing left upon his body, and his upturned face, by which one could identify the perpetrator of that wicked deed.

Holmes was determined to be a man and being no longer able to help any of the wreckers, he turned his attention to his own business. He had had the goods insured in St. Louis, and proposed that we sell the insurance to the merchant of whom we must get our supplies. This we did, and on the seventeenth of April we started out from St. Joseph. Our first six miles west from the river, which we crossed on a ferry, was through a lightly timbered section. We then struck a prairie, upon which we found a little belt of timber along its streams. The soil or mould, was black, loose, and fertile.

Taking a north-western course from St. Joseph, after crossing several small streams, and for some time following the Little Blue, we came to one of its tributaries which was too deep for fording. The water was at that time unusually high; we must, therefore, cross it in some other way than by wading. We had prepared ourselves for such emergencies. We had made blocks to put on the bolsters of the wagons and under the boxes, to raise them nearly to the tops of the stakes. Having put these blocks upon the several wagons, we tied the wagon boxes down so that they could not be washed off by the current. By this means we crossed two streams before reaching Fort Kearney.

By this time several small companies, going our way, had joined with ours, making our number about sixty-five. We did not hold ourselves responsible for their safe passage through the western wilds; but we traveled together, and were all by agreement subject to certain rules. We were fast approaching a country where it behooved us to

> 'Be up and watching,
> With a heart for any fate.'

We therefore put out a guard every night to watch the camp. This guard was appointed by a daily draft, or better this, that every man took his turn, which was determined by response to roll call, and he was obliged to serve or furnish a substitute. There were generally two on picket together, and in cases of evident necessity, more were appointed, as circumstances dictated. We traveled at the rate of eighteen miles a day through this section of the country, and were now where we did not see a woodsman's clearing, or a prairie home. We were "Out West," that uncivilized, indefinite, out of the world sort of place, which so many people never expected to see who now live there, in many cases even in opulence, and who still say "Going out West," as though there were no "anguish west count.ee" east of them. We were beyond the clamor of Presidential campaigns, and coming elections, and cared nothing practically for all their filibustering, if the people would only make sure of having the District of Columbia well stocked with good official stuff. So long as the American Eagle, and the Star Spangled Banner should wave over our heads, and the red man should not pick the pockets of this man of his tobacco, nor of that one of his last "mon," and the coyotes should not pick the bones of any of us, we counted ourselves happy, and richer by far than any rascal in the land who should go to the polls to deposit, with a telling effect, a pocket full of votes.

The antelope began to show themselves and I began to show signs of increased happiness. Sometimes I would see them on the road ahead of us, or not far from the side of the road as we passed along. Of course I used to take my rifle and go out after them. No sooner would I start than three or four, or a half dozen of the men, taking their guns too, would follow me. Knowing the excessive timidity of this most beautiful creature, I soon resolved to go quietly back to the teams, when any of the men should follow me. So many hunters would always scare away the game. It was, therefore, arranged that I should do the hunting for the party, and as a compensation should be released from picket duty.

When about sixteen miles east of Ft. Kearney we pitched our next camp. We ate our breakfast very early in the morning, and as we were finishing our meal an Indian came up from the brush that grew close by the river bank, came into our camp and wanted to know which was our chief. We pointed him to Captain Nutter. Stepping up to him he asked, "You chief?" "Yes." Striking his hand upon his breast, the Indian said, "Me chief, too; good Indian." By this time another had come in, presently a third, and in a few minutes a "big heap Indians" had come into our camp, all of them mounted upon ponies or mules, and all armed with guns. There might have been three hundred of them. The Captain was frightened. If any of you were witnesses of the childish, yet actual fright of the little school girl, who, in her blissful ignorance of the fact that there were any colored people, was so terrified at the entrance of a colored boy into her school that she could neither study nor recite; that after the children had learned what was the matter with her, and had set him up to all sorts of mischievous pranks, and she used to hide in every hidable place; that when she stood near him one day in the spelling class, and he conscious of her fear of him had fixed his white-black eyes upon her, and she was so overcome by it that the teacher, supposing her to be sick, sent her home to be cared for; if you saw her a few months afterwards, when in a new home she was one day visiting a family in whose house there was a colored servant, and when that servant, suddenly making her appearance from the basement, entered the living room, she could not a first, for her fright, utter a sound; but, her whole frame in agitation, she started and in an amusing attitude stood pointing her finger towards the black visage; and when this new mode of treatment so amused Aunt Dinah that she opened her great white eye-balls and stretched her mouth to the fullest extent of its great capabilities and thus brought the child's struggling words to a squeaking utterance, she cried, "There, see! see there! there!" If, I say, any of you knew of this girl's fright, you are easily prepared to appreciate the Captain's condition on this memorable morning. We did not tell you at the first all that we knew about our friend Nutter. He was sometimes badly frightened. Rushing to the cooks he said "Stop, don't; don't stop to wash the dishes. We are in a bad spot; we shall all be killed." Holmes and I, just before the Indians had come up, had for the safety of the guns shot them off. Says Nutter: "Put up those guns; they will think we are afraid of them." "They will know we are ready for them," I said. One of them came up to Holmes, or the Colonel as we often called him, and wanted him to give him some tobacco. Holmes took out a large piece from his pocket, and cutting off a small piece gave it to the Indian. That made him angry, and in an instant he sprang to grasp a sheathed knife from Holmes which was attached to a belt and hung at his side. The Colonel was a large, powerful man, but quick and nervous, a man who was afraid of nothing; there was not one cowardly trait in his disposition. I've sometimes wondered if the Devil himself could frighten him much. With his powerful fist he struck the Indian, and set him standing on his back. These Indians belonged to the Pawnee tribe. They were the largest, most muscular and most powerful Indians I have ever seen, and in fact this is true of the tribe as compared with all other tribes. Some of our men were terribly frightened, and began to give them everything they wanted. One man from Virginia, who had with him his family, and who belonged to one of the companies that had recently joined ours, had two teams, and from one of the wagons he began to feed them with crackers. I said to him, " Don't you give them any more ; you will starve to death." " What shall I do ?" he said. " Give them powder and balls ; I am good for two of them at the first shot." One of the Indians looked up at me and laughed. I have always thought he understood what I said, and I read in his laugh that they were not after our blood, but our victuals. It proved to be the correct translation of their purpose. Finding they could not get much they left us. We got up our teams and started on.

There were two brothers from Missouri who had joined us. I have forgotten their names, if in fact I ever knew them, for we always called them Missouri. They had been across the plains in 1850. They had with them two cows and two horses which they were going to take to California. One of these cows had a little calf. These men were so badly frightened by the Pawnees that they did not even look for their cow. The Indians stopped about two miles beyond our camp. We passed them on our way, and when we were about two miles beyond them I missed the cow. I spoke to one of the men about it and found that they already knew that it was left behind, and that they were afraid to go back after it. I offered to go with one of them to get it. He was afraid to go unless one other man would accompany us. We returned to our camp, where we found the calf. Its throat had been cut and it was left to die. On our return to our ox train we found the cow in the hands of another party. When we reached Ft. Kearney we found our Pawnee friends there and found the officers rolling out to them barrels of flour and pork. They had told the government officers that they were about starved, that they had intended to rob our company; but they found some very resolute men among them. The officers were driven to give them something. The Sioux Indians were then at war with the Pawnees. We went on a little beyond the Fort and camped for the night.

The next day we went about as far as usual and camped by the river again. The river was at this place about a mile wide, but exceedingly shallow; it was at no place more than two feet deep, and was full of sand bars and islands. This sand is cold quick-sand. We hitched six yoke of oxen to our several wagons and crossed the river. We could at no place, while crossing, stop our teams lest they should sink in the sand. The day had passed its middle mark before we were all safely across the river. We were now on its northern bank. We stopped for the afternoon to rest ourselves and our cattle.

After dinner some of us thought it a good way to rest, to go out hunting. The Colonel, one of the Missouri men and myself started, followed in a short time by fifteen or twenty others. Back from the river flats we could see buffalo. The country was very level for four or five miles from the river. Having crossed this level tract, we came into an undulating region, a beautiful rolling prairie country. The bluffs were not high, but they were treeless and almost shrubless, smooth, clean bluff upon bluff, over which the eye could reach in its clear vision mile after mile. The tops of these bluffs, as far as the eye could reach were covered with herds of buffalo. A grand, good sight for my eyes. Missouri said that he had killed a great many buffaloes, and that one must hit them farther back and lower upon the body than any other animal. Agreeing to aim as he directed, I soon after fired and hit one of the animals; but I knew by its motions that it had been hit too far back. I was sorry; still I believed that I could, by carefully changing my position, kill it. That I might do it, Missouri agreed to keep perfectly quiet; but at the critical moment he called to his brother, who was about eighty rods behind us. My buffalo, which wasn't mine, called to its brothers, and the whole herd went off. The Colonel and I were, as you may suppose, indignant that we had thus lost our game; but by telling of him, I have now had my revenge. We sat down and beheld the grand retreat. The land as far as the eye could reach was blackened with the huge, wild creatures. There were doubtless thousands of them, and somewhere in their number the wounded one.

We turned our course and soon saw a herd of twenty-five or thirty coming towards us. We trid to head them off, but we were too far away. We shot at them, but did not kill any. There had been so many men out amongst them that day, that they were excited and easily frightened. We started for camp. On our way we saw a buffalo coming, followed by a man on horseback, driving at full speed. When we had come within a hundred rods of them, the buffalo, doubtless frightened by us, turned about. Its pursuer was now only a few rods distant and he fired at it. He hit it; but after running a little ways, the two being in close proximity, the buffalo started to fight the man. He fired at it again, and the creature fell and very soon died. The man had shot it in the heart. As we helped him turn it over I made up my mind that I had shot my buffalo too far back upon the body. I resolved, and I think it an advisable resolution for any one to make, that I would always shoot as close to the fore leg as possible, and about one-third of the way up the body. As there is a hump on the shoulders, unless one is careful he will shoot too high. We all reached camp in good season; but I cannot say that we did in as good spirits, for none of us had brought into camp any game. I found at night that I had left my powder-horn, doubtless where I had made my first shot.

We had planned to get an early start in the morning, and to make a big drive. We therefore started before sunrise. I quietly asked Hibbard to go with me to get my powder-horn, and suggested that while we were gone we might possibly get some game. We told no one except Holmes where we were going, lest the fright among the wild herds should be repeated, and we again defeated, and our men as bad off as villagers without a meat market. As the teams started we left the road and took our march towards the bluffs. I found my horn; but there were no buffaloes in sight. We moved on westward, and after a little struck for some bluffs a mile or two from us. As we reached the top of a bluff we saw large herds of buffaloes, two or three miles away. Our courage was good and our determination to push our way on towards them. When we were within a half-mile of them, there seemed to be two ravines, one on each side of them. I said to Hibbard if he would approach them by the left hand ravine I would go up on the right side of them, and we might both get a shot at them. When I had reached a point in the ravine which I thought was opposite the herd, I carefully ascended the knoll. I saw one get up. I fired at him. He ran a little ways, then fell dead. Just as I had reloaded my gun, three antelope ran past me; aiming at one of them, I shot it, and when it had run on a few rods it tumbled over. I approached it, and unjointing the hip bone I carried the saddles off with me. I cut out twelve or fifteen pounds of the buffalo steak, and putting my gun barrel through the skin, which, by the way, I had left on for this purpose, I carried it over my shoulder. I could not find Hibbard, so I went on alone.

I thought I was about ten miles north of the road. I started in a southwesterly direction. After a time I saw several elk, and a little further along I saw a buffalo go down into a ravine. I approached, on my way, within five rods of it before it saw me. When it did it was frightened and ran off. I was willing it should go. On, and still further on I went. I crossed a large flat piece of land, beyond which there was quite a formidable ridge. As I looked upon it, and with my eye picked my way to its summit, I saw five Indians standing near the top of it. I looked upon them and I thought they looked at me. I thought it rather mean in them to get so exactly in my way as they had done, but resolved to go on and take my chances, and meet my fate with a brave heart. We were then in the Sioux country, not exactly in Tartary, but where good Tartars lived. As they saw me coming they seated themselves as though ready and waiting for me. Well! I thought, I'm ready for you; yes, and good for two of you any day. As I neared them, I found they were those red—y men I had not expected to meet. My Indians proved to be friendly ones. They were Hibbard and four of our men, who left the train soon after we had started out in the morning. I was glad to find they were white men, whom I knew. They had killed no game, and could help me carry mine. About two o'clock we came in sight of a small herd of buffaloes. The "boys" wanted me to shoot one of them. I crawled up in shooting distance of them, fired at one and broke his shoulder. He started and ran down the hill in a direction partly towards the boys. They ran and surrounded him and commenced a series of firings upon him. I sat down to see the fun. They shot ten times, each man shooting twice. Chase advanced upon his victim, which was now in a hollow or sink hole. I called out, "Don't go any nearer; you'll get hurt." But he replied, "I'm not afraid; he is most dead; there's a stream of blood running out of his mouth." Buffalo, upon this depreciation of his ability to fight, put down his head, scraped the ground with his fore foot, and made one desperate leap for Chase. He was then about twenty feet from the buffalo. The former started in hot haste; the latter followed close upon the rear. It was a run for life, and nearly even, too. The buffalo ran about fifteen rods and laid him down to die. Chase stopped as soon as he learned the fate of his pursuer. Of all the yelling and screaming and jumping I have ever heard, theirs, during this little scene, was amongst the loudest, and withal the most ludicrous. Such scenes are not often paraded upon public or private stages: this scene was natural, and it was wild. The men were really, literally, actually frightened out of their senses for a little while; but for the bystander, who had advised a different course, yet who believed in the end they would come out unharmed, it was decidedly laughable. They cut off what meat they could carry. I took up my load and we went on, thankful for our fun, and as thankful that none of us had been hurt. We had had nothing to eat since sunrise: we had not found any water; but despite hunger and thirst we pushed our way onward with resolution. We had lost the road, but were determined to find it. We were very tired : one thought we ought to leave our meat; but the others thought it best to keep it for future use, in case we should ever get a chance to eat again.

Just at dark we came to a brook. We could see, by their lights, thirty or forty Indians in camp not far from us. It was too dark for them to see us. With renewed courage we pushed on, not to their camp, but away from it, and our courage was doubtless increased by the hope of getting off without their knowledge of us. I believe we did the best that men could do, in picking our way along through the darkness, and in a strange country. In about an hour we struck the road. But here a new difficulty arose: we did not know whether our company had yet passed this point, whether they were in advance of us, or still behind us. The probabilities were, that in our circuitous marches upon our hunting grounds we had lost time, and that our men were westward of us. We resolved to go on until we should come to some camp. We fired our guns, but received no response. We traveled on another hour, then fired our guns again, and this time fire gave answer to fire. I need not tell you that we were joyful. 'Twas good news to us. Hoping that the shot came from our camp, we marched on with new life. Two of the men from camp were sent out to meet us. They proved to be our own men. We reached camp about eleven o'clock, exchanged joyful greetings, and recitals of the day's adventures. They feared we had been captured or killed. I had never been so thirsty as I was that day. The only water we had found was at the brook where we saw the Indians. I had tasted of it; but it was strong alkali water, and not fit to drink; what I did drink made me sick. The next day I was not able to sit up.

We had had no fresh meat to eat since leaving St. Joseph. The buffalo meat was dry and tough, the antelope's tender and good. The men ate very heartily of it, and one young man became very sick. It was Davenport's boy who was sick. He grew worse for several days, and died. We had stopped our train during his sickness, and did everything for him that it was possible for us to do. Mr. Knapp preached his funeral sermon, and we laid him away in the best box that we could make, and drew a large flat stone and put it over his grave to keep the coyotes from digging him up. While we were in the midst of the funeral services the coyotes were on a knoll about sixty rods from us, fighting and howling so dismally that it was difficult to hear the preacher. They had doubtless scented the corpse, and were in angry waiting for a chance to tear it to pieces.

A few mornings after, some buffaloes crossed the river, and came towards us. The men began to cry, "Cartwright, Cartwright! get out your gun, here's some game." I had resolved not to shoot another buffalo while crossing the plains, lest some one would be made sick by eating the unsavory, unhealthy stuff. I told them no. Then some of the men wanted the horses, and to go and hunt the buffaloes themselves. As there had been an agreement before starting that the hunters should not have the use of the horses, we denied them their request. Nutter and Johnson took the two mules of Virginia, as we called our Virginia comrade of whom we spoke at the time of our morning call from the Pawnees They started on their hunt, after a while got separated from their party, ran three or four miles, a buffalo turned upon them to fight them: they shot it several times, and at last succeeded in killing it. After this supply of buffalo meat had been devoured, the men were satisfied. and did not again ask for any more of such meat.

We were still following the Platte river on its north side. and for two hundred miles we saw no trees. There was just one tree that stood upon an island, but it was dead, and there was only one limb upon it. It had doubtless been stripped for firewood. We were obliged to do as were other travelers across the plains, to use buffalo chips for our fires. We made ditches about eighteen inches wide, and building our fires in them, we would then put our kettles across the ditch.

The Platte river, because of its sand bars, its rapidity, and its shallowness, is not navigable ; though it is wide. The valley through which it courses is remarkable for its length, and for its fertility in the eastern portion of it. Its width is from eight to fifteen miles. Having passed this section of the valley, and following up the north fork, we were fast approaching the "bad lands" which lay between us and the Black Hills. The soil was sandy, and was beginning to grow alkaline in its character. Grass was becoming scarce. The land back from the river was covered with a small brush called greasewood, and with wild sage The latter seems very much like wormwood. The prickly pear, or the cactus, grew in abundance. Acres upon acres are still covered with them. Buffaloes were becoming scarce, and antelopes were numerous. The company voted me clear from all other duties to hunt, and I furnished our own company, and those that had joined ours, with all the fresh meat, except one deer and one antelope, which was eaten upon the remainder of the journey. After getting an early breakfast I used to start on in advance of the teams, kill my antelope, drag it to the road, and go on. The men would pick it up as they reached the spot. Everything went off nicely when the weather was favorable, and we did not drive so far as to tire out the men. Sometimes we would be obliged to drive until late to get where we could find water and grass.

Our road was very good until we came to the Black Hills. The scenery on either side of our road was monotonous. We were hemmed in by bluffs, which shut out from view the more picturesque landscape beyond, and we saw little for a long distance but the bottom lands, treeless. springless flats. The region known as the "bad lands," and close to the hills "is about thirty by ninety miles in extent, sunk away from its prairie surroundings with almost vertical sides, and is about three hundred feet deep in its lowest part. It is filled with innumerable pinnacles, columns, and irregular masses of earth and rock, separated by labyrinthine passages, nearly destitute of vegetation, bare and sterile, but rich in fossils, geological treasures, and organic relics of extinct animals."

For many miles along this section of the river the atmosphere is so clear that the extent of vision is almost incredible. When we were opposite Chimney Rocks we were about five miles from the river, and about twenty-five or thirty miles from the rocks; yet we could distinctly see them, and could also trace the outlines of trees as they stood against them. One man told me that he had traveled the road on the south side of the Platte, and that after seeing the rocks he traveled more than half a day to reach them, and gave it up. They stand probably fifty feet high, and bear a strong resemblance to an assemblage of old chimneys. I saw teams, men, and one day an antelope at the distance of eight miles. A man at a distance of three miles would seem to be ten or twelve feet tall. We usually traveled sixteen or eighteen miles, and yet, while here, we could at night look back to our camping ground of the previous night.

About seventy-five miles of our journey through the Black Hills was very mountainous, and in places as poor. The road led back from the Platte, because the mountains came, in many places, abruptly to the river. One day there seemed to be a city some six or eight miles west of us. It proved to be the case; for the place is known as Rock City; but the city, though built of stone houses at a comfortable distance apart, is yet uninhabited. The rocks and piles of stone are, at a distance, quite house-like in appearance.

When we reached Ft. Laramie we traded off some of our sore-footed oxen. A man was there whose business it was to exchange cattle with travelers. Two days out from the Fort we saw an immense drove of cattle on the road. Of course we wished ourselves ahead of them, and Nutter commenced a race. We drove for two days and one night, stopping in that time only long enough to cook and eat our victuals. Any detailed description of our table manners upon the road during this hurried march, would be, if in keeping with the meals themselves, so soon given that you might think that we ate nothing. But we did. Suffice it to say that the meals were as enjoyable as all the circumstances in the case would allow, or as any that the young folks take when the old folks are away from home. There was more jollity than formality. We were on a spree, were not exactly on dress parade, nor on exhibition; for there were no wayside spectators. Some of the jolliest times we had on the route were during those two days. I have not, in the twenty-two years, forgotten to laugh at some of the ridiculous performances of that cattle race. The horses and cattle of the opposing party were fresh and strong, full of spirits, and running as if for life. We were to pass through barren lands, and the foremost party of course stood the better chance of finding water and grass. We won the race, and our reward was therefore the better feed.

We soon struck the Sweet Water river. To the left of the road, as we neared this stream, we saw a famous rock, which in its coloring resembles the pipe stone rock. Hundreds of names are inscribed upon it, truly indicative of the "Young America" spirit, calling for honor and renown. Two miles from this rock there is a small alkaline lake. The deposits of alkali which are formed when the lake dries up in the summer, are in some places of a considerable depth. Wagon loads of it may be, and are, then gathered with little difficulty. It is the common pearlash or soda, of commerce.

The Colonel and myself went out to the lake, and on nearing it we saw a huge buffalo wolf a few rods from us. He was sitting on his haunches, and looking at us. Having foolishly left our guns with the teams, we could only do the best thing that circumstances might dictate. We picked up a stone and threw it at him. He greeted its harsh reception by rising in a decidedly snarling mood. His only attempt at smiling at the honor of a call from white men was an extended stretch of the mouth, with a marvelous display of white, ugly teeth. To meet this stranger any more than half way, was to look into the jaws of death, if, indeed, not to enter the same. We had no idea of traveling in that direction. If we had seriously attempted it, our purpose would have been thwarted, for at every step of ours, which brought us nearer him, that we might, if possible, stone him to death, like the coward that he was he retreated. It was a long time, however, before he left us. There was but one thing that prevented his feasting on the sweets of our flesh, and that was his natural cowardice in the presence of an antagonist, and the extremity of cowardice in the presence of a fearless human being. Many persons who live near the habitations of wolves, or who travel through their countries, would sleep better at night, and rest better by day if they would allow themselves to believe the truth of this statement, that wolves, though ferocious animals, are sneaks and cowards, and rarely attack men. Their howls make the night hideous, and their distorted features make one cringe; but, belonging to the dog family, as they do, the saying that "barking dogs never bite" is applicable to them. Even now I regret that I left my gun in camp that day. Such an impertinent staring at as we received at the hands, or, more literally, from the eyes and the jaws of that wolf, should not have gone unpunished. If we had only had those guns that day, we would have issued his death warrant from our gun barrels, and with our balls would have executed it upon his luckless head.

We followed the north side of the Sweet Water for twenty miles, then crossed it. At the point of crossing there was a ferry. The boat was small, and it was by a very tedious process that teams were conveyed across the river. There were a number of companies at the ferry, each one waiting its turn to cross. At the rate of passage, we would be obliged to wait three or four days before we could move on. True to his instincts, Nutter planned a way of escape for us from such tedious waiting. The river was high, but he believed we could ford it. We raised our wagon boxes as we had done before, and fastening ropes to the lower side of each box, drew it over the top, and with some men on the up stream side to hold the ropes and keep the wagons from turning as we struck the current, we started in. All of the men were taken over on horseback. The first one who went over carried a rope, one end of which was fastened to the foremost yoke of oxen. The oxen were hitched to the wagons as when upon the road. There were men on horseback on the down stream side to whip up the cattle as they struck the deep water. Our risky undertaking came out all right, and in half a day we were safely crossed to the further bank of the river.

One of the men in the crowd of those waiting a transport, having seen us cross, thought to follow suit. Having taken no precaution for the safe passage of his wagon boxes, when his teams struck the current, despite his best efforts at that late moment, they were turned first down stream, and then over into it. He lost almost all of his supplies, and it was with the utmost difficulty that he kept his team from drowning. He had been advised not to drive in as he did; but he was one of your self-willed men, who know their own business, and whose success in life is the measure of the soundness of their judgment.

The road on the south side of the Sweet Water was level and good. It was a well traveled road, and probably as good as any country thoroughfare in any of our States. Soon after crossing the river we reached Devil's Gate. The ridge of rocks which lies on both sides of the river averages about three hundred feet in height, and extends nearly north and south, reaching about a mile on the north side of the river, and thirty or forty rods on the south side. As we first observed it, it seemed like one solid rock, possibly five hundred feet wide. On the south side it slopes gradully from the river to the ground. We reached it early in the morning and camped for a few hours to look at it more carefully. As we ascended the rock, we found on reaching the river that there was a clean cut passage for the flow of the water. The stream as it passes through the rocks is fifteen or twenty feet wide. The rocky wall, as we looked over it into the river, is neither perpendicular nor perfectly smooth; but there are here and there projecting crags, many of which at our distance from them looked tiny, but could we have been near them, and they were in keeping with the general structure of the rocks, would have looked massive. Could we have seen the man who thought that the Natural Bridge must have been built by the Devil, and could have shown him this wonderful gateway, we presume he would have said that the post holes had been dug, the posts made and set by the same dignified personage. We did not think that, and were not unimpressed with the grandeur of the scene, and did not leave it without an increased reverence for the great architect of the universe. On a crag about one hundred and fifty feet from the top of the rock, there lay the body of a man. How, or when, or why he had reached that spot, none of us knew; if he had been murdered by a white man or a red, if he had accidentally fallen over, or had willfully thrown himself over, were equally undecidable questions. One of the hands had in some way become unjointed, and lay at the surface close to us. It had doubtless been carried there by some vulture-like creature: the flesh had been picked off, evidently by some bird. Mr. Knapp carried it to California; but what afterwards became of it I do not know.

Our road through the Sweet Water valley lay the most of the time close to the river, and it was generally good until we reached South Pass. At that place we reached the highest point on our route, the elevation being seven thousand four hundred and eighty-nine feet above the sea. The atmosphere at this height was so rare that it was difficult for ourselves and our cattle to breathe, and doubly tiresome to march. The combination of varied scenery gave us a grand picture. There is a notch in the snow-capped mountain, a beginning of vegetation lower down, and a gradation of it, until, when we have looked to the foot of the grand old mountains, we could see a growth of heavy timber, magnificent in its growth and venerable in its age. The valley is perhaps three-quarters of a mile wide. The night that we reached the Pass, we camped by the Pacific Springs, and there first saw water that flows into the Pacific Ocean.

We camped earlier than usual to look about. The most of the men took a tramp from the camp, some going in one direction, and some in another, some to look at the valley, and some to look upon and from the mountains. Some of the party saw elk and some saw mountain sheep. Several of the men who had been out together returned to camp in great excitement. They were not frightened, but wild with joy. They had reached a California good enough for them. They had found gold in large quantities. They had filled their pockets with the precious dust, to show to us poor fellows who had been so unfortunate as not to learn the fact so soon as they did. But unfortunately for them, their gold proved to be but mica, a yellow isinglass.

Our first camping ground west of South Pass was sandy, upon the Little Sandy creek, which empties into the Green river. We were there obliged to fill our water-sacks with water. These sacks were made of rubber and would hold about twenty-five gallons. They were to be tied at the top like a grain sack. Our next day's tramp was a long and wearisome one. We marched about twenty miles to reach the Big Sandy, which also empties into the Green river. I say we marched; for that we always did. No man could ride, unless unable to walk.

Nutter rode on horseback in advance of the company to look up camping grounds. The cooks followed him, to facilitate the cooking affairs.

At the Big Sandy we again filled our water-sacks; for we would find no water between any of the streams through this section of the country. Our cattle and ourselves too were obliged to drink the water carried in the sacks. As we could have none any better, we were obliged to make the best of our condition, taking what little satisfaction we could from the fact that all travelers upon the road must needs share as poor a fate; but after all, we were really not so joyous over the fact that misery had its company, as appreciative of the sufferings of many who were not well provided with the means of transporting water, and whose sufferings were in many cases very distressing. Twenty-eight miles lie between the Big Sandy and the Green river. This we were obliged to make in our next march, or camp where we could find no water. The weather was very warm, it being about the first of June. We traveled at a slow rate, making only two miles an hour.

While eating our dinner we saw an antelope coming up from a rise of ground. The Colonel went onto a knoll near by, to attract its attention, while I went onto the other side of it to shoot it. I crawled through the sage brush, until I came as near to it as I wished to, then waited while the Colonel drew out a handkerchief, which he had tied to a stick, and waved it. When the antelope saw him, it started and before it had made its first circle, in its attempt to approach the object of its fright, it came within shooting distance of me, and I killed it. It was a beautiful specimen of the antelopian family, and furnished us with very palatable meat.

At sunset we reached the Green river. It is not more

than ten rods wide, but is deep and swift. The road brought us to a ferry, by means of which we must cross the clear, cold, and beautiful waters of the stream. The rate of ferryage was very high. We paid twenty-five cents for every man, one dollar for every horse, one dollar for a yoke of oxen, and five dollars for every wagon. The boat was run by ropes and pulleys. The ferrymen were taking in, in those days of gold fever, from three hundred to five hundred dollars a day. We camped over night on the eastern bank of the river and were obliged to wait until noon before our turn should come to cross it. The last company to cross before we could go was a small one: it had packed mules. As one of the men was getting on his mule it jumped: the man was thrown off and his foot caught in the stirrup, and as the mule started off he was dragged along with his head upon the ground. The mule made a circle about twenty rods from the river, kicking at almost every step. When it reached the river side of its second circle, it plunged off into the stream, ran about half way across it, and then made a very short turn to come back. The mule was in deep water when it turned, and the man, whose foot was still held by the stirrup, was out of sight. He was probably kicked as the mule turned about; just then he threw his arm up out of the water, then disappeared forever. We took boats and tried to find him, but could not.

That night we camped by a beautiful little stream beyond the Green river. We had here the prettiest camping ground of the entire route. The mountain stream, for such it was, was a tiny thing, but clear, sparkling, and beautiful as it rippled over its stony bottom. The valley on either side of it is not wide, but wide enough to furnish us ample room for ourselves, our stock and our goods. The hills immediately adjacent to the stream are of moderate height, smooth, and as beautiful to look upon as are any to be found. Hills of greater height rise back of them. The scenery is not wild; but for one of serene, joyous beauty it is a gem. In the morning we went on three or four miles and camped on the same stream. It was our rule that we should not camp two nights in the same place, though every seventh day we would rest ourselves and our teams.

We found a great many wolves in this part of the country. We saw the big buffalo wolves, the black wolves and the gray ones, and the coyotes or prairie wolves. Those last named were the most plenty, and were the noisiest of them all. One night one came into the camp and carried off our kettle cover. This kettle was really our tea kettle, a dish made of sheet iron, one that would hold twelve quarts. It had a tin cover. This the wolf carried off, and we found it outside of the camp, badly bruised with the angry gnawings it had received. One day I saw a very large white or buffalo wolf, and four black ones following it. What was the intention of these black, fiendish creatures was more than I could tell. The large one was either trotting or galloping along: the black ones, following close upon the rear, would snap at him; but as soon as he would turn upon them they would retreat. When he would turn around and run on, they would follow suit, and would snap at him, and bite him again. They went around a knoll and were out of sight. I went onto the knoll, but could not see them. While there looking for them, I saw an antelope; but it was beyond shot. I secreted myself behind some grease wood and raised a handkerchief. The antelope saw the handkerchief: it ran in a circle about me; then it made a series of half circles, with every half circle coming nearer me, until when it had come within shot of me, I fired and killed it.

The demoralizing influence of such a life as men led while crossing the continent as they did in those years, when the accommodations for travel were so poor, when men must pass through the countries of so many hostile tribes of the red man, and when the excitement for gold-digging almost crazed hundreds of men, the demoralizing influences, I say, were only with the greatest care resisted. Men leading a sort of nomadic life, cooking, washing, mending, doing everything for themselves, sleeping out of doors night after night, walking seldom less than several miles and often many almost every day for months, many times regardless of the weather, traveling as best they could over unbridged streams and mountain roads, seeing no white men upon the road except those traveling like themselves, either east or west, and the ferry-men and government officers at a few points upon the way, surrounded at night by the wild beasts in their wild homes, followed by day by the same blood-thirsty creatures, and upon the greater portion of the route subject to a sudden attack from the Indians, such men often became strangely reckless. They became reckless of their health, of their manners, of their morals, of the comfort of their fellow-travelers, and reckless of life itself. They often displayed that extremity of recklessness that by its very demoralization still evinces the superior workmanship of man's Creator, beings created in the image of God, but who by the development of their baser natures may become fit only to populate a hell.

Two men who had been neighbors in Connecticut doubled their teams and traveled together. They had four mules and two wagons. They carried two young men with them. After a long time the owners of the teams got into a dispute, which grew hot and then hotter. One wanted to go faster than the other. The latter proposed that they divide their goods and teams, and let each man take his own time. They did so, and the former went on at his desired fast rate. The latter, taking his march more moderately, overtook him on the fourth day. He had stopped at the foot of a hill worn out, and his beasts unable to draw their load up the steep road. As he saw his acquaintance approach him he asked him if he would hitch his team on to his own and help him draw his wagon up the hill. The man said "no," that he had all he could do to get his own team through, and told him wherein he had failed, that he should have taken the advice not to go so fast. It was the old, old story of "I told you so." The slow man started his team up the hill. The fast man asked if he wasn't going to help him. He said "no." The fast man then stepped to his own wagon, took out his gun, walked to his neighbor, and in the presence of the two young men, killed him instantly. He then stayed by the young men, not even offering to escape from them.

When the next company reached them the young men told of him. The new-made grave also testified of his guilt. They carried him with them about three miles beyond, where they camped for the night by a stream. Two or three companies following stopped there with them. They kept him in custody, and at night they formed a jury, and appointed a judge. The two young men were the witnesses. He was pronounced guilty and sentenced to be shot. Blanks were drawn and three men who should draw them were to shoot him. We reached the place the night following the morning of his execution. The two young men were still there, having stopped with the ferrymen. The prices charged at this ferry were also exorbitant. Again, through the management of our Captain, we succeeded in fording the stream, rather than to pay the price, and be so long waiting our turn. We went forty or fifty rods down stream and crossed it in safety to ourselves and our goods, by such management as on a previous occasion already described. The young men took the teams that had fallen into their hands, and traveled with us until we reached Steamboat Springs, on Bear river.

A few days after this we overtook a woman who was sitting on a wagon tongue. She was entirely alone. Of course we heard her story; for we were anxious to know why she should be thus left alone, and in such a place. She said her husband's cattle had drunk so much alkaline water that they were sick, and were all going to die. He was watching them. His brother who was traveling with them, was discouraged, and unwilling to share their fate with them, or even to help them out of their trouble, had gone on and left them. She was moaning and bewailing her lot, and begged us to kill her. Her distress had so overcome her that she was anxious to be put out of her misery, even by facing a rifle shot. She was so crazed that she was unaccountable for her words or her wishes. We could do but little for her. She could not leave her husband, and he could not then leave his cattle. We overtook her brother-in-law about four miles from where we found her, and prevailed upon him to go back and help them.

We were at this time traveling through the country of the Crow Indians. They were friendly, and we were glad. Our cattle not being so carefully watched while we were traveling amongst the Crows, strayed from camp one night, and in the morning when we went out to find them we came upon a camp of the Indians, and they went out with us, to help us find them. Well, why couldn't all the Indians be "good Indians," and not keep us in such a state of excitement, as they often did? Our travels through the Crow country were not unmarked with pleasurable events; but, as you cannot turn your attention from those events, possibly uninteresting to you, and actually behold the country through which we passed, nor its inhabitants, it may be well to sum up the matter by saying that our journey through this section of the country was in some respects unusually pleasing. It gave us rest from those anxieties that were upon us when we made such long, hard marches by day, and watched so sharp by night, when among hostile red men. Possibly some of the boys would like to know that for twelve sucessive days I killed an antelope. The day before we reached Bear River mountain I saw an antelope, and asked Hibbard to go with me to shoot it. He went on beyond the animal, intending to fire when all should be right for it. The wind was blowing so fiercely that he could not stand still, and he feared that the motion would frighten the animal. He therefore shot, but the wind miscarried the ball. The antelope, in its fright at the shot, turned towards me, and I killed it. We were about two miles from the road.

Before reaching it, we came upon a party of Indians, forty or fifty in number. Their camp was near us. They were all mounted on horses. They stopped and saluted us, and performed for our benefit, or our amusement, I cannot say which, their war maneuvers. They would lean far over upon one side of their horses, as if they would hide behind them, and would bend their bows as if shooting upon an enemy, from under the necks of their ponies. They divided into two companies, to show us what they could do, and how they would do it, were they in an earnest fight. They had war clubs hanging upon one arm, in which there were notches, varying in number according to the number of those whom they had scalped. One of them had a club upon which he proudly showed us seventeen notches. When we reached our company, which had, meantime, gone on some little distance, and had camped for dinner, we found the Indians already there. They there performed their feats again, and I may say to the pleasure of all. It was pleasant to meet friendly Indians in a far western wild home, and to see them acting according to their own customs. They were en route for Steamboat Springs, or the Bear, or Soda Springs, as they are also called.

That afternoon we doubled our teams to cross the Bear River Mountain. It is so very steep that it is impossible to cross it by any ordinary driving. It was the steepest, and almost the highest mountain crossing on our road. We hitched six yoke of oxen to a wagon, and by dint of perseverance we succeeded in reaching the summit. We then left our wagons, and went back after the others. The road on the east side of the mountain was possibly two miles long. To get our teams down the mountain we were obliged to take off five yoke of oxen, to tie the four wheels together, to put a chain from the forward wheel over the wagon box, and have two men at the rear to hold onto the chain, to have two men to hold down each upper wheel, and a man in front to hold the teams and keep them from going too fast. We dispensed with drivers for the time being. None of these precautions were unnecessary, for the road was so very steep. The road was possibly three-quarters of a mile long on the western slope of the mountain. This road was a new one, and very rough and poor; yet it was a decided improvement upon the old one crossing the mountain: formerly travelers had been obliged to take their wagons to pieces to get them over in safety.

The Bear River valley is very beautiful. The Soda Springs produces blood warm soda water, which, when it is sweetened, tastes like our soda water summer drink. Three miles beyond it, we saw the Steamboat Spring, otherwise called Windmill Rock, and Old Crater. The rock from which the water issues is about four feet high, and four across it, and is nearly round. There is a hole in the center six inches in diameter. The water comes through this hole, and is forced about fifteen feet high. It shoots as if forced by wind or steam, and makes a noise, as it issues from the stone, like that of a high pressure engine in a steamboat. Water springs from the rock about twice in a minute. It is like the water of the Soda Springs, but not quite so strong in its alkaline quality. We reached Steamboat Spring in the morning, and remained there until the following morning.

We found our friends of the previous day already there, and, besides them, others. There might have been a thousand of them. There were six tribes met in council. They were very friendly, full of fun, a jolly, good-natured set of men. We enjoyed our afternoon spent with those Indians, and think of it as one of the brightest spots upon the road. If you say that all Indians are lazy, thievish, treacherous, I am still as sure as though no one doubted it, that these men were as friendly, genial, and manly, in all that we saw of them, as any one could well demand of any person. It was well worth the few hours that we waited, to see a bona-fide Indian council, assembled in so great numbers, and upon their own grounds, talking and acting in their own natural way.

Some of you lovers of horses, would not have been at a loss for enjoyment. Many of their ponies were very fine, and their races were beyond anything in that line that I have ever witnessed. You who would not be known to attend a horse fair, and you who delight in the race, would have been stupid, if you had not enjoyed this sight. Fine horses, a great many of them, expert drivers and as many of them, and the matchless races! They came about us and asked us if we had any race horses, and if we would bet with them on their horses.

As we took our march from Steamboat Spring, we left the Ft. Hall road, and took the California road to strike the Humboldt. On this road, which was in good condition, we found plenty of water, because we crossed so many creeks running into the Snake river. The feed was good and abundant.

One day we found a two-headed snake. A part of the company was in advance of the rest of it, by some little distance, and finding the snake as they passed along they killed it as they supposed. But quite like the poor rule that doesn't work both ways, it is a poor two-headed snake that can't work itself both ways, and as its two heads were really better than one, it had commenced to crawl away from its murderers. When we who were behind reached the spot we found it still alive. It was a little thing: it measured sixteen inches in length: in size it was at one end about like a lead-pencil, the other end being larger. There were two perfectly developed heads, but the one at the tail end was the smaller one. It could crawl as well one way as the other. If a stick were put in front of either head it would back away from it.

The Digger Indians lived along the Humboldt river. They are the lowest, most degraded, filthiest beings of their race. I have no doubt that they will do as it is said they will, dig up the bodies of dead men and eat them. As we came into their country at the head of the Humboldt, we found traces of their murderous raids. We found graves of men whom they had murdered: there were head-boards at their graves upon which were marked the dates of their murder. We found one new-made grave. The man buried therein had been shot only a few days before, while on guard. We kept a double watch at night while in their country.

The Humboldt is a narrow stream, which runs down in sinks, and empties into a lake of the same name. It runs over a sandy soil and is always riley. We followed it for a long, long distance, on the north side of it. We went over a sandy, desolate plain, a fit abode of such carrion-like creatures as the Diggers, and yet it is more than possible that the climatic influences, and the almost complete destitution of vegetable or animal food are strong impelling influences which drive them to manslaughter for the preservation of their lives. There were willows along the river banks, and in some places the clusters were dense. By the bends of the streams there were patches of excellent grass.

It was some time before we saw any of the Diggers; but we had reason to believe that they stealthily watched our camp almost every night. In the morning we would find their moccasin tracks ten or twelve rods from the camp. One night while on a bend in the river, some of the men shot off their guns. They happened to fire into one of the clumps of willows, and five Indians who were secreted there came out and ran across the river in the shortest time imaginable. Had they been running after us, it would not have been so laughable: as it was, it was about the funniest thing of the sort I've ever seen. They were badly frightened and ran for life. One morning we came upon the camping ground of a party of white men, which was then two or three miles in advance of us. One of their oxen had died the night before, and the Indians so soon after their leaving the camp had carried off the most of the meat. We saw none of them while passing through the ground; but as we reached the top of a hill, but a short distance beyond, we looked back and saw several of them emerging from the willows. One very bright night I was with four others on guard. I thought on such a night I should surely see the Indians, if they did come about us. Our cattle had been turned out for grazing upon one of the grassy spots at a bend in the river. I lay all night on the ground by the side of an ox, that was also lying down; but I did not see one of them. Again, in the morning, we found their tracks not more than ten rods from us.

The 4th of July found us near where we left the Humboldt. We stopped our march in the middle of the afternoon to celebrate the day as best we could. We knew that the inhabitants of the country were not in full sympathy with such a movement. All we asked of them, however, was to let us alone. We fired our guns, first, simultaneously, then in rotation, and we got up the best supper that could be provided. Uncle Sam has never complained of us for not doing better that day.

Not far from South Pass we met a company of twenty-five returning from Yreka County, California. On leaving the big bend of the Humboldt, we turned from the main California road which led directly to San Francisco. We took a north-westerly course and followed the directions of this party. They described a road which had been traveled in 1849, and which led directly to Sacramento. We followed it for one hundred miles, then made a road of our own, only as in places we followed the trail of this party. On those places which I have called their trail, there had never been a white man's track except theirs.

We left the Humboldt in the afternoon, purposing to travel by night while we crossed the alkaline desert of the Humboldt Valley. This valley, so called, is, however, only a section of the Great Basin, the comparatively level connecting land between two of the lofty ranges of the grand Rocky Mountain system. Through this valley, as also through north-western California, and southeastern Oregon there are many evidences of geological disturbances. Many of the hills, devoid of vegetation, are thickly covered over with a crumbled mass of rocks and stones of varying sizes, evidently the work of volcanic action: springs of warm, of hot, and of boiling water are common, and extensive tracts—of land can I call it?—of alkaline salts spread out before the eye. These fields of glistening salt are at first sight beautiful to look upon. The atmosphere, which over this section of the country is exceedingly clear, lends enchantment to the view, an extra polish to such desert sand; but to the weary west-bound traveler, to whom a pleasant change of scenery would be a rest, the monotony of such a desolate desert becomes tiresome and depressing to his spirit. Besides, the glare of the white desert waste is blinding. It is, therefore, unsafe for men to travel over them by daylight.

The first morning after leaving the Humboldt, we reached a spring of very warm water. A beautiful little stream made out from this spring, and close by it we found a patch of good grass. We made our camp there for the day. Upon leaving it we came upon the same white, barren desert, and all along this road, for a distance of eighteen miles, we found the remains of terrible destructions of camps, of all that pertains to them, of men, of cattle, of wagons, and of other camp and traveling furnishings. We found cattle dried up, the flesh shriveled, but nowhere broken, the hair perfectly preserved. We found where wagons had been left standing, and in some cases the wooden parts had been burned, the earth and air being so exceedingly hot here, nothing but the wheels were left. Some of the wagons were in a perfect state of preservation; the chains were not unhitched from them, and in one case the cattle had turned the wagon a little from the road, and had then laid down and died.

Early on the following morning we reached the Boiling Spring. It is at the foot of a very high mountain, and measures about twelve feet across it. The mountain sides were covered with small burnt stones. There was not a tree, nor shrub, nor spear of grass to be seen anywhere upon the mountain. The spring evinced the still angry elements under Dame Nature's control, the stone-capped mountain the traces of past anger, of changes that were doubtless slow, but that in themselves were grand.

The water of the spring was, as the name indicates, boiling hot. We dipped water from it, and steeped our tea in it without other boiling. What could be the harm in that? It is said that fish are sometimes found in boiling springs, and that from their motions they do not seem to be away from home, or out of their sphere. They must be boiled fish. It must then be all right to boil one's tea in such a spring. Afterwards some of the men washed clothing with this hot water, and tying ropes to the garments threw them into the spring to be boiled. The fastidious need to share a little of the rough experiences contingent upon such a journey, or, if they will not, we refuse to hear their reproaches for such a method of procedure. Besides, we found here a combination cook stove and boiler, which, so far as we knew, had not been patented. As we were all lovers of filthy lucre, we wished to know what good purpose such an invention might serve, and if possible some money might be made out of it. Of this I am certain, we would have earned some credit, if not cash, had we soon after this day's labor met any jaded and dirty looking fellows from the west.

From the Boiling Spring a good sized stream runs, but settles into the ground about a mile further on. We found good grass for some distance along the stream. As we had twenty-eight miles to travel before we should find water again, we took a vote of the company, getting one majority in favor of starting on at four o'clock in the afternoon. We had traveled all night for the two nights previous, and as it was so hot all along the road we could not sleep during the day : we were, therefore, very tired, and many of us were not fit to march on. There was no chance for any of the men to ride, unless one should be too sick to keep up. Our cattle had all they could do to carry our stuff. Rogers and myself gave out. We put our traps into a wagon, and lay down upon our warm spring bed. These springs were a stange invention, they would spring down, but not up : it was more like what I might, from its softness, call a feather bed. We determined to sleep, at all events. As it began to grow light we took up our march, and very soon found two others who had given out by the way. Every two or three miles we found some of the party along the road. Among the stragglers we found an old man whom we always called Indiana. He had urged the forward march, and as we came up to him he said, "Go on, I've sent on my vote for the march, whether I ever catch up or not." We overtook our teams in the middle of the forenoon. They had kept up until six in the morning. We rested with them until the next morning.

By taking this northirn route we had less than thirty miles of travel across the desert between watering places. The ordinary route would have obliged us to go sixty or seventy miles without water.

Thirty miles beyond the desert our road lay for six miles in a canon. A stream ten or twelve feet wide found its way between the two lofty mountains. There was possibly a distance of six rods between the mountain on one side and the stream on the other side of this deep, dark passage. All along perpendicular walls of rock loomed up on either side, varying in height from one hundred to four hundred feet. In some places the rocky walls must have been four hundred feet for a long distance. Our road lay over very rough and stony ground. The stones were large, and it was with difficulty that we could drive. It was emphatically a hard road for man or beast. Sometimes the pass was so narrow that we were obliged to drive through the water; but this we could do, for the stream was shallow.

We camped one night in this narrow bed. At our camping spot, it would hardly do to call it ground, there was a cave, the entrance to which was four feet high and six wide; the interior measured eighteen by twenty-two feet, and the height at the center was twelve feet. In this cave a dozen of us slept. In the canon we found several barrels of whiskey that had been left here in 1849. The Indians had found them, and had opened them and taken out several gallons from one barrel. The whiskey had spoiled in every barrel except one. Some of our men lived, to their shame be it said, high and fast that night, and found themselves "tight" before they got out of the place. We found a blacksmith's vice and an anvil in the cave. They were new, but slightly rusted.

When we emerged from this canon we found some timber, and were not again out of sight of timber. We soon saw game. I saw some antelope on one side of the road. I went out and shot one, and put the saddles upon my back and followed up the company. We had had no fresh meat for a couple of weeks.

Just after I reached the road I heard some noise behind me. On turning about to learn its source, I saw a wolf not more than a dozen rods off. I laid down my venison to attend to his wants, as he had without doubt called me. But as I turned towards him, he turned his back upon me. He didn't want me, after all; so I went on. Again he growled and followed me: again I turned upon him, and he as soon from me. Both times when he left me he hid in some sage brush. I did not see him after his second hiding.

About this time we struck a spur of the Sierra Nevada, and at its base before us lay a beautiful valley. In this valley we found the finest clover one could set eyes upon. It was like our common red clover, but of uncommon growth. It was gotten up on the California principle of doing everything on a large scale. As we walked through it, the tops of it would reach the shoulders of our tall men. Some of the blossoms were white, but were as large as the red ones. We also found our common white clover: the blossoms and the leaves were small, but the stalks grew as high as the red clover stalks. The most of it was lodged. It was excellent wild feed.

As we were going up this valley we saw a lake, and some antelope at the left of the lake. The Colonel, who was, by the way, one of the bravest men I have ever known, went with me to get one of them. Soon after we started we saw not far ahead of us several Indians. The Indians here were hostile. I said to Mr. Holmes, having first spied them, "Shall we turn about and go to our company, or go on and meet them?" He stopped a moment, ripped out an oath (for I've heard him do such a thing when under great excitement, or when determination was fired by desperation), and said: "No! we'll go on and meet them. I'm good for three of them, and I know you are for two." He counted five and had them all provided for. If the Colonel felt any fear, I could not with the most careful scrutiny detect it. Why we should have escaped harm at the hands of the Digger Indians and of some others whom we had met, I cannot tell, even now. The fact that we had so many times escaped from their vengeful hands was no evidence that we should at this time: it really only lessened the probabilities in our favor. I confess my mind was full of queryings as we neared them. But we did then, as we had always tried to do when in the presence of hostile Indians, we met them with a boldness that covered our fears, yet was not ostentatious. Marching thus through their midst, we managed as circumstances at the time dictated, and came off unscalped, unharmed, and were undisturbed only by the weakening influence of the relaxation of our fears, as we saw on reaching them that they were five of our own men. We had supposed that the road would take us to the right of the lake: having, then, no reason to suppose that there were white men near us, except in our company which we had left behind us, our imagination claimed dominion over us and a right to say that the men were Indians, and no doubt of it. The road, however, led to the left of the lake.

We camped for the night by a spring at the base of a very high mountain. This was one of our specially beautiful camping grounds. There were lofty, magnificent pine trees upon the mountain whose tops were almost out of sight. The trunks were large and clear of branches for a long ways up from the ground, thus affording a clear, clean passage underneath. Could a few acres of these pine trees be transplanted to some prairie of the

Western States, or even to the mountainous East, they would become the wonder and the admiration of the country round and the enviable picnic ground of any community. The valley was covered with a luxurient growth of grass. It seemed as though nature must have designed it for the habitation of a more civilized, more appreciative people than such Indians as lived there.

But who shall say that the Indian who knows every stream and dell, every mountain and mountain pass of his country, has no appreciation of the beauties of nature? Some of them were, to say no more of it, sadly behind their privileges if they did not love to look upon many of the sections through which we passed. Could this place have been inhabited by intelligent white men, I could have been easily satisfied to spend the remainder of my days within its precincts.

In the morning we commenced the ascent of the mountains. I say mountains; for there were a series of them, or of foot hills, as the lower ones are called, every next one higher than the one we were then ascending. We reached the summit of the mountains in the afternoon. Holmes and I left the teams, to search for game. We were going slowly down the mountains, when I saw to the left of me (now this was not an Indian scare) a very nice buck standing behind a log. I could see about one-third of his body above the log. I fired at him. The Colonel had not seen it, and inquired what I did that for. The animal ran on a few rods and fell. After drawing it to the road we left it for the men to pick up when they should come along.

We went down to the valley and camped for the night. By that camp I saw the largest bear's track that I have ever seen. It measured twelve and a half inches in length and seven in width. We there found a fallen pine tree measuring seventy-five paces. We were not amongst the giants of the forest, the big trees of California, and are not telling stories to see which can beat, we in telling, or you in believing.

In the morning I started early to find game. I went further down the valley and saw six antelope feeding. I went up around them and got in ahead of them. When they came up within fire I shot one. The others ran towards the camp, and the men seeing them, spread themselves, unnoticed by the timid creatures until they saw themselves fairly surrounded. They did not wait long, however, before starting to run between the men. Several shot at them; but none of them were hit, and fortunately none of the men were hurt, though they had been shooting towards each other.

That day we began to see signs of the Indians, their tracks, their fires. The next day they began to build fires close to the road, sometimes on one side and sometimes on the other; but they themselves kept in advance of us, so that we did not see them. The third day after crossing the mountain we left the old road, the one traveled in 1840. We crossed a stream which runs into Sacramento, and taking a north-westerly course, we had nothing to guide us but the trail of the party which we had met at South Pass. We stopped for dinner by the stream, and afterwards when we had got upon the flat we saw two Indians coming towards us, and about half a mile from us. They were as wild as any deer. When they saw us, as they doubtless did, they ran off into the woods. The side hills and mountains were covered with timber; but there was no timber on the flats. We camped by a very pretty little stream. The flat was as pretty. On the opposite side of the stream some rocks rose perpendicularly to the height of one hundred feet.

During the night the Indians came up onto these rocks and rolled off stones: we supposed with the intent to scare our cattle and raise a stampede amongst them in the night. We penned in our cattle as well as possible by our wagons, and we put out a double guard. No disturbance arose; for our cattle, though frightened, were in close quarters. The day following we saw a great many tracks of the Indians, but we saw none of them. There seemed to have been great numbers of them. The trail was quite well worn, in places where the ground was hard. In sandy places, where the tracks could be more easily observed, they had taken some brush-wood and drawn over the tracks to obliterate them. We were somewhat puzzled to know what their object could be. They meant something by it, that we knew, and as the Indians were hostile, it meant something besides a "Welcome Englishmen," such as greeted the ears of our puritan fathers after they had come upon the eastern shore of our United States.

We camped by the side of a lake, marked on some maps Goose Lake, on others Grove Lake. It lies in California, except its northern extremity, which is in Oregon. That night we lost our trail. The Indians had intended that we should lose it. We knew there was something to pay, and that soon.

We were in the Modoc country; and now you who have never seen a red man's trail, you who have never seen them in their native wilds, have never seen them upon their war path, nor heard their fiendish war-whoops, nor their diabolical yells at a scalp dance, are still well prepared to believe the worst that could happen to us, while among the Modocs. You have not forgotten the cunning and the duplicity which they served upon our Peace Commissioners, not yet two years ago. You have not forgotten for how long time they succeeded by treachery to outwit and to out fight the troops stationed upon their ground, nor have you forgotten the final victory of our men, and the sentence pronounced upon the ring-leaders of the Modoc tribe.

It is said that twenty years ago they were powerful, and were engaged in warring against the white man. I do not wish to corroborate the statement—I should hardly feel justified in doing so—knowing as I do by my personal knowledge of them, three years previous to that time, that they were only glad to fight white men who passed through their country. We did wish to find our trail, and we did feel fully justified in making our best efforts for that purpose.

In the morning we sent out four men to find it. Two of them were to go out from the lake in a north-easterly direction, and two were to follow the lake shore upon its eastern bank. The two going out from the lake came, at the distance of a mile, to a reef of rocks which followed the lake for a long distance. In some places there were several rods between the rocks and the lake, in others only a few feet, being only just room enough for a team to squeeze its way through. The South Pass party had described the lake to us; but the Indians had fooled us by wearing the tracks which we were to take. When the men got beyond us a mile or so they found the mules' tracks, and came back, reporting that we were on the right path. They did not see any Indians while away from the camp. Taking their direction we started on, expecting to overtake the other men along the shore of the lake. We were about to enter the narrow pass between the rocks and the lake when our other men came up. They had followed the lake until they came to a mass of rocks along which there was such a narrow passage

between them and the lake, that they went to the right of
the rocks. There was a deep projection of the land into
the lake, and in the center of it there was a reef of rocks
about half a mile long. As they reached the north-east
point of the rocks, following a short distance upon a trail,
they supposed it to be the path leading to a stream at the
head of the lake, which the South Pass party had told us
we must cross. They told us that we were on the wrong
trail. The rocks at our right ranged from fifty to two
hundred feet in height. This reef was not a solid mass
of rocks, but a loosely packed mass of smaller rocks, with
deep fissures here and there. To the imaginative, or the
timid person, or even to the practical literalist, the crevices in those rocks furnished capital hiding places for an
enemy, and there were those in the company who were
glad that they were not obliged to go on that way. There
were grasses and reeds growing along the edge of the
lake. They grew eight or ten feet high above the surface
of the water, and in some places were very thick. Some
of the men, prompted by an irresistible curiosity to see
what they could, and some determined to learn if there
were Indians close by, began to scale the rocks. They
spied them in the crevices of the rocks, and in the grasses,
and called out that the ground was full of them. We
formed a breast-work of our wagons, fired off our guns
and got everything in readiness for an attack from the
Indians, having scarcely a doubt that they would come
upon us. When we were ready we started to go where
the last two men had told us we should find the road.
The Colonel and myself had each a good revolver, and
feeling a measure of responsibility for the safety of the
men, we felt that we must take the lead, and be the first
to face the danger. As soon as the Indians saw us go
down by another trail, for they were slyly watching, they

started quick, passing between us and the lake, to cut us off, as we supposed. But, as " Every road leads to the end of the world," and as going to California seemed like going to the end of the world, we were bound to take that road. The men whom we had sent out to find the road had not been far down upon the trail, therefore none of us knew that the Indians had a camp in ahead of us. When we reached the ground, their camp fires were still burning; but their ugly, howling dogs were the only living testifiers of their camp quarters. They had failed to break our lines, as they had doubtless supposed they would do when we entered the narrow pass, and seeing us start towards their camp, thought we were going to deal out vengeance upon their squaws and papooses, and ran to notify them of the coming danger, and to clear the camp. They had secreted themselves in the rocks which filled in the greater portion of the point of land, and were hidden in the reeds and rushes growing in the lake. The most of them were secreted in the rocks, and yet, as we ascended these, we did not see them. We saw that the rocks ran out into the water at the point, and that the trail led no further than the camp. The camp flat was about two rods wide. The rocks were from seventy-five to one hundred feet high. As we returned to follow up the other trail, which we now knew must be the right one, they came out from their hiding places like a swarm of bees. We knew they could sting, too, and we were not professional bee tamers. Two of the men were with me at the rear, driving up the loose cattle, when the teams started to go back. Several Indians came toward us with their hands uplifted and palms open, as if to say they wouldn't hurt us, they were weaponless. One of the men wanted me to wait and see what they would do. I did not know whether it was Captain Jack and his four or five braves who stood there before us. I didn't know his name in those days; but I suspected their duplicity, and preferred to go on. As I was no Peace Commissioner, and would not furnish the Captain any chance to shoot me in the back while making any treaty or parleying with them at their call, it was doubtless well for me that I did go on. As I looked off to my left I saw forty or fifty running in the grass, bent over to secrete themselves, and evidently intending to cut us off from our party. I said to Mr. Cole, "See there! those Indians are trying to cut us off." He raised his gun as if to shoot them and they ran into a clump of trees close by. As they emerged from them they presented a formidable array of bows and arrows ready for effective work. I called to the company to wait for us. They did, and we saved ourselves a second time from the clutches of the Indians. When they found they could not catch us as they had hoped to, they jumped into their canoes, and put straight across to the head of the lake. Its eastern shore, along which we passed, was convex, and they therefore gained rapidly upon us. When we reached the point of the lake where they were, they allowed us to pass. It was a surprise which we accepted with gratitude. It was near night when we passed them. The rocks near the head of the lake ran off to the east, giving us a pass of about fifty rods. We followed the trail described to us, and at eleven o'clock we reached the stream, running into the lake, at a point where there was a natural ford. The Indians followed us up, keeping a half mile in the rear. Again we were surprised; for they did not molest us during the night. It was a moon-light night, a circumstance in our favor. In the morning we found that they had camped about three-quarters of a mile from us. We crossed the stream and saw them no more.

The fifth day after our experience with the Modocs one of our lame cattle strayed from camp. The men were unable to find it. At sunset, just as we went into camp, we heard a gun about a mile on our back track; but we paid no attention to it. In the morning nine men overtook us. They had found the lost ox. On leaving the Humboldt we had put up a sign-board. Upon it we described our party and the road we purposed to take. These nine men, having read our notice, had followed our track. They came with packed mules, and traveling so much faster than we could, the Indians were not apprised of their coming, and knew nothing of it until the company emerged from the pass at the northern end of the lake. The Indians at once surrounded them. The men broke through, but left their mules and all their provisions. For four days they were without food, except what berries they had found along the way. When they came upon our noon camping ground, they found the lost ox and our fire still burning. To lose no time, and by all means not to lose us, they had determined to drive the ox and hurry into camp. Night overtook them and they were not able to travel further. They killed the ox. It was that shot which we had heard. When they reached us in the morning we gave them, not exactly hasty pudding, but because of their necessities, a hasty meal. They kept with us during the remainder of the journey.

That night we camped under Mt. Shasta. Its snow-topped peak had been a good guide to us for many a mile. Since we came in sight of Pike's Peak, we had not been for a whole day at a time out of sight of snow. We had also not been for so long a time without seeing emigrant trains until after we left the Humboldt. When we were in those countries where we could look back upon our road for a long distance, the trains seemed to be so close to each other as to give the appearance of long, and quite compact processions. It is probably true, as some have estimated, that the thousands of men who traveled across the plains during 1852 and for several years, might have been counted by the scores.

Nutter went into Yreka and we went on a part of the way. In the morning a party was sent out from Yreka to ask us to camp at a place six miles from the town. They wanted a chance to kill the fatted calf, in honor of our arrival with the first immigrant teams that had ever been driven into the place. On the third day from Mt. Shasta we drove in and were treated by white men, like white men. The banquet was very creditably prepared, and there was such a sound of revelry that night, the 7th of August, as we had not heard for many weary months. "and all went merry as a marriage bell;" — but, in a few days there came that deep sound, which struck like a rising knell, and we heard it.

A man who had belonged to a party of ten came into Yreka. He alone, of his company, had escaped the murderous raids of those villainous Modocs. They were surrounded by them before they knew that there were Indians in the vicinity. Nine were at once killed. He broke through and followed our trail until he reached the ford. He supposed they were on his heels. His mule gave out, a strange thing for a tough creature like a mule to do, but a tougher and more mulish trick for the creature to give out at this time of the man's extreme necessity. However, the mule gave out, and the man took to his heels. He traveled all night and just at day-break he came back, having traveled in a circle, to where he left his mule. He found it refreshed by sleep and food: he mounted it and came on the trail, reaching Yreka without other trouble than that of hunger.

On hearing this, a party of about eighty was formed to go back to Goose Lake, and give the Modocs what they deserved, a thrashing with a gun-barrel for a flail. Capt. Nutter went as guide. A Yreka man went as Commander-in-chief. An Oregon Indian who was generally known as "Oregon" and who had before been out with such parties, also accompanied them. Unfortunately for my peace of mind, I was unable to go with this party, as I was at the time sick.

They returned, reporting that they had killed fifty. They said that when they reached the lake and came upon the leaders, the "Capt. Jack" of that day, came out from the point of rocks which secreted their camp, ran around as if to warn his fellows, meantime shooting his arrows into their midst. One of the men killed him. When they saw that their Chief was dead they were badly frightened and went pell-mell to find places of safety. Many of them hid in the rocks. Capt. Nutter found two squaws trying to hide themselves in some crevices which were a little too small for them. They could not turn about in them, so they had crawled in feet first, and he said, "I could see them looking out at me, and it seemed savage in me to shoot them; but I suppose it was right." He shot them. The squaws in the camp started with their papooses to cross the lake. Oregon caught a canoe and put after them, killing and drowning as many as he could. When Capt. Nutter asked him why he did that, he coolly replied, "Nits make lice." They took one prisoner. They told him to show them where the rest of the Indians were, for he professed to know where they were going to camp. Oregon could talk with the Modocs, so he said to the prisoner, "If you will show us where the others are, you shall be released, but if you fool us you shall be killed." He did fool them. He took them to a perfectly barren place, one destitute of Indians, or anything better. He said, "I am mistaken, they are over yonder." in such a place. The men told him he should have but one more chance of that kind. Again he deceived them, and Oregon, stepping boldly up to him, struck him through the heart with a knife, saying as he did it, " You shall never lie to me again."

As the Yreka company first neared the Indians they found a party of sixty or sixty-five that had been surrounded by them. None of them had been killed, for they were so thoroughly barricaded by their teams. They were hemmed in by rocks, and cut off from the water. They were in a pitiable condition. There were two women in the party: one of them was an elderly woman. She was sitting close to one of the wagons, and holding onto an axe with the desperation of despair. It was with great difficulty that they could induce her to yield her grasp of it and receive help at their hands. She was almost, if not really crazed by exposure, and fatigue, and fright She intended to use the axe for self-defense, and if worse should come to worst, to swing it right and left in a general fight. This company was released and sent on. Our company also found fourteen dead bodies, which were mangled and terribly butchered, lying near the lake. If they were a part of the company to which the one man belonged, of whom we have already spoken, or if they composed another company none of the men could tell, but probably the latter. They buried the most of them. Some they could not bury.

Having brought the company into Yreka, and in safety, Mr. Holmes, Mr. Nutter and myself had filled our obligations, and the men dispersed, some going in one direction, and some in another. We had been on the road a few days less than five months.

Several of us went onto Green Horn creek, three miles, from Yreka, and took up a miner's claim. We worked about four weeks. Being inexperienced in mining, we found at the expiration of that time that we had made but thirteen dollars apiece. Torry, Hibbard and myself went onto Scott's Bar. The mountain was covered with sugar pine. On the western slope we were obliged to go down around a point by going through a rocky channel which was just wide enough for our mules to pass through. Everything must be packed to be conveyed over this mountain. I saw several mules that had fallen over deep precipices. There were rocky walls of four or five hundred feet in height. As we reached the first settlement, we passed a prospect hole. It was four and a half feet across the top, and twenty-two feet deep. While Mr. Hibbard was on his mule, and had just passed the hole, the mule began to back, and back it would. It fell. The horse and rider lay at the bottom of the hole. The prospect for them was not so fine as the retrospect. When the dust had cleared away so that we could see what to do, we got them out of it. Davenport, our fat friend, was the Samson of the occasion. Hibbard was wounded on the head, and for some little time showed no signs of life. He was kindly cared for, for several weeks, by a man who was a stranger to all of us, but who would at no time take any pay for his services. I hired out to work in the mines for a few days, and meantime lived in a vacated camp. The man who had previously camped there came and dug up some gold which he had secreted beneath the ground. From three hundred to five hundred dollars were taken out of this mine daily. I bought a claim, for which I was to pay the first three hundred dollars which the mine should produce, and returned to Yreka.

Mr. Holmes then returned with me to Scott's Bar. When we were upon the summit of the mountain, I looked to the right of me and saw a deer standing and looking at me. I shot, but overshot. My pride was a little touched; as I had never before missed my mark when the Colonel and I were out together. I reached for his gun, took it and killed the deer. We went to work on a claim, and the first day we got two hundred dollars; but the gold soon run out, and there was not enough to pay for the claim. I was not in perfect sympathy with the business. We went to hunting. We could get twenty-two cents per pound for our venison. We cleaned our guns for hunting, and the next day were ready to start out.

A little Indian came into our camp, affirming that he belonged to John's tribe. He would take one of the guns and shoot with me at a mark. We shot them off several times, and he then asked to stay all night with us. We had but one bunk in our tent. There was a Mr. Babcock living close by us in a shanty. He invited the Indian to stay with him. In the morning Mr. Holmes was obliged to go off with a prospecting party. I started out to hunt. The Indian, learning what I was going to do, wanted to take Mr. Holmes' gun and go with me. He consented to let him take it, and we started out together. We expected to find our deer on the second rise of ground as we ascended the mountain. Having reached this rise of ground, there was on one side of us a ravine, or gulch. We could not look onto the side of it nearest us; but we could look across upon the other side. We therefore decided that I should go across and should shoot the game on his side of the gulch, and he should shoot that on my side. He was to remain where he was until I should reach the other side. The underbrush grew eight or ten feet high. I went over as agreed, and stopped first to see where the

Indian was. I could not see him, and at once suspected he had run away with the gun. I heard something behind me, and turned just in time to see him leveling his gun to fire at me. Not such a friendly Indian, after all. 'Twas a new phase in the rough experiences of a woodsman. Quick as a flash he dropped his gun when he saw me looking at him. I think if he had had the courage thus to dispose of himself, he would have thrown himself over into the ravine, and never have looked at me again. I called him up and told him, as if I meant it, clat-a-wa, or go ahead. He did until we reached the top of the mountain, when we came upon several deer, and there he killed a very large buck. We took it to camp and dressed it, and I gave him a part of it and told him to go home. But he wanted to stay all night with us. He commenced to make a spear head of the deer's horn, alleging that he was going to spear salmon trout. We left the camp for a little time, and could scarcely have been out of sight when he stole our guns, powder, and powder horns, our lead and caps. Our ball moulds were in a hollow log, and he had found and taken them too. We very soon learned that he belonged to the Klamath tribe. They were hostile. The Mr. Davis who told us offered to go with us to find our rifles. The Klamaths were fifteen miles away. We learned on reaching them that he had not been seen for about ten days. Soon after this the Chief of the John's tribe went for us and found the Indian who had stolen our rifles; but he refused to give them up. Mr. Davis sent word to the Klamaths, if they did not return the rifles he would take men enough with him and they would exterminate their tribe. He had previously fought them, and as they stood in fear of him they gave over the rifles. The Chief took forty-five dollars for his trouble.

A few days after this Mr. Tuttle, from Massachusetts, wanted me to go hunting with him. It was near night. We saw no deer, but saw a great many signs of bears. We followed a stream that flowed past our camp, and that started from the mountain where I had before hunted. There was thick brush close to the stream. The first hill was very steep and bore evident marks of ancient volcanic action. We could easily follow a bear's track through the crumbled stones. As we neared some thick brush on our return to camp, Mr. Tuttle was a few rods in advance of me. He had seen grizzly bears, and turned to me saying that he should never fire at a grizzly bear, if he should see one. I said I would fire. Possibly a spirit of combativeness and a little of braggadocio spirit was the prompter of this remark, as I felt safe at the time to make the assertion, knowing as I did that I had no grizzly bear to shoot, and that probably Mr. Tuttle would not be there to see me run from it, should I ever meet one. But no sooner had I made the bold assertion than we heard a crackling noise in the brush, followed by a stranger noise, and in an instant we saw a huge grizzly bear coming up from the other side of the creek, with a cub at her side. She ran up a few rods above the brush, then stopped. The cub stopped too. I shot. What else could I do? My honor was at stake, whether my life was or not; the momentary surprise and (shall I say it?) humiliation forced me to the shot. As the ball struck the bear she took her paw and struck the cub, by the act throwing it down the mountain. The old bear ran towards the brush, then stopped to look at us. Tuttle was nearer the bear than I was and he ran. I stopped him—told him to "wait and see a big fight; I was ready for it." Oh! what a brave boy I was with my big plum pudding; but, like the little one, I wanted some one to see me, else there was no fun in it and I imagine there would have been only a one-sided fight.

I do not mean to say that were I alone in the woods I would necessarily be a coward; but that many times it has been true in my case, as I also believe it has been in yours, that when I have been in positions of danger if there were those with me who were frightened, especially if frightened beyond self-control, my courage has risen, and always in proportion to the emergency. The practical use of a little courage gives more to the one who is actuated by it, and it keeps those from falling who are weakened by their fears. In the present case it may be that the best thing I could do was to fire at the bear; possibly it was not. But having been caught in a trap of my own setting, I could do no better than the most cunning animals that I had ever entrapped, I could but make the best possible use of my powers for self-protection as circumstances at that late hour might dictate.

I knew that a wounded grizzly bear was a doubly formidable antagonist. Tuttle said she would kill us. My gun was ready. She stood a moment, then she and her cub ran up the opposite hill. Tuttle fired at her, but did not hit her. It was useless for me to fire at her in her position at the time. She had gone up onto the hill and on about half a mile. It soon seemed as though she had rolled down the hill. She was bleeding profusely. I saw her going into the creek, where she was very soon out of sight. I thought I would get in ahead of her. In a moment she came out square against me, and about twelve rods from me. Again she struck her cub and drove it off sixty rods upon a foot hill. She then returned — growling, and snarling, and breaking the whistle wood as she came. Tuttle started to run, saying that she would call all the bears in the neighborhood. Once more I stopped him and told him to stand by and see the biggest fight he had ever seen. She returned to the spot where I had shot her before, and we went through the same maneuvres. As she stood looking at me I shot at her, but did not hit her. She walked off and acted as though she were dead. I then started to kill her cub, but had only started when the old bear stood up on her hind legs, and began pacing towards me, growling at every step. I fired again, and again I missed her. She went up the creek a little ways, and lay down by the side of a large balsam. I went up on the side hill to shoot her in the head. Just as I was ready to fire, I heard a noise up the hill, and on turning to learn the cause of it, I saw a bear coming, jumping and snorting as it came. It ran within six rods of me. That was too much for me. I had not bargained for that, neither had I been trying to shoot all the bears in the neighborhood into our presence, but rather to shoot those already there out of the ability to touch us. We both started for an oak tree, which we could climb and thought the bears could not, and, indeed, we supposed that a grizzly bear would never attempt to climb a tree. We were not fully posted in bearisms. The third bear did not notice us. It had come at the call of the old and the young one to the help of the latter. The mother bear had thrown her cub down, doubtless, with the intent for it to call for help, and thus released from the care of it she might give us a little attention by way of one of her affectionate bear squeezes. The cub was pacified, and as the third one had, probably, not seen us at all, they went off in peace, and gave us a good chance to kill the wounded bear and make good our own escape. I had both times shot the bear through the lungs; still it lived nearly an hour. Had it been a deer, and thus shot, it would not have lived long enough to run twenty rods. The day following, five of us took it to camp.

Poor health made it necessary for me to return home, and to take the easier route. Passing through the Golden Gate, I went by way of Panama, Aspinwall, Jamaica, and New York to my home in Wisconsin.

VII.

HUNTING TRIPS IN MINNESOTA AND NORTH-WESTERN WISCONSIN.

A TRIP IN NORTH-WESTERN WISCONSIN.

Mr. L. B. Green and myself started out from Madison, Wisconsin,— to which place we had gone by rail,— traveling on foot; for we were in search of government lands. The second day out we crossed the Wisconsin river. We went through Sauk City, past Devil's Lake, through Baraboo and on to Webster's prairie, where we stopped for the night. I was taken sick and was obliged to remain there several days. Mr. Putnam and his son-in-law, at whose house we had stayed, took their team and went on with us. I was not able to get in and out of the wagon alone, and had I not been used to living a good share of the time out of doors, and at disadvantageous rates, it would not have been safe for me to start as I did.

During the day it commenced to rain very hard, and we asked an Irishman whom we chanced to meet, how far it was to the Ohio House. He thought a moment and said: "Just half a mile from where you now stand." But we drove five or six miles and having found no Ohio House, we asked another Irishman whom we met how far we were from it. He thought a moment too, — they seem to be very thoughtful creatures, these Irishmen,—then told us, "Just seven miles from where you now stand." On we went, distrusting strangers and having within ourselves little ground of confidence that we should ever see the house in question. We rode a half mile and found it. There we stopped for the night. The next night we camped out, and keeping my feet almost in the fire I burned out a good share of my rheumatism. The following night we put up at the Globe House, in Sparta. The house was not finished, but there were a good many guests there. In the night a cry of "Fire!" brought every man to his feet at the same moment. The fire proved to be in an adjacent building and not our own, as we at first supposed, and the only loss that the travelers sustained was the common loss of boot-straps, which were found in the morning thickly strewn upon the floors.

After we had passed the Black River Falls, we stopped at night at a private house, where we found that we were needed to help take care of a stranger who had lost an arm by the accidental discharge of his gun. Being near the house, he crept in and was cared for there. The wolves had tracked the blood on his way to the house, and the man would doubtless have been devoured by them, had not the inmates of the house happened to hear him just in time to save him.

The country through which we traveled was poor: the soil was light and sandy, and the timber very scarce. We went on and into a poorer country than this. I thought it could be put to one good use, namely; to colonize snakes. Snakes were already so abundant there, that I believed the climate and the soil both well adapted to their perfect development. We saw some massasaugers while there, and some bull snakes — large, spotted ones. One was six feet long and another was still longer. A few miles northwest of this immediate section we came into a good country again. When we were about ten miles beyond Menomonee we passed the mail carrier. His horses were hitched to a tree, and he was lying on the ground asleep. At first we thought we would hide the mail bags and then wake him. We did not do it; for policy demanded that we keep on the good side of Uncle Sam, if we expected him to give us all a farm, where sand and snakes were not so thoroughly intermixed as to make poor soil for good seed, such as we intended to sow.

When we had made our camp for the night, I started out in search of deer. Just at dusk I saw, about twenty rods from me, something black, quite bear-like in its appearance. I fired at it and it fell. I loaded my gun and started towards it. As I neared it, the animal raised up and commenced biting a tree. I intended to shoot it through the head; but as I fired the bear dropped its head, and the ball went into the tree. My next shot killed it. Just then I heard something in a little tree close by me. I looked up and saw three tiny cubs. I called to the men to bring the axe; for I wanted to get them alive. Having heard me shoot three times in so quick succession, and then yell so loud, they thought some Indians were after me, and before coming to answer to my call, they made some preparations for their own safety. On learning what I wanted, one cut the tree down, while two of us stood ready to catch the little fellows as they would fall. Putnam's son caught one and I caught two. His bear began to bite him and he called for help. One of the men took one of mine, so that I could take Van Estian's, but by some mismanagement it got away. Another one was soon lost, but mine I kept until it became quite attached to me, so much so that it would howl as soon as I was out of its sight. I sold my little bear on board a boat, shortly before I returned to my home.

IN THE CHIPPEWA REGIONS IN WISCONSIN.

A few months after my trip into the Chippewa regions in 1856, William Lee and my son Jonathan accompanied me into the woods about Menomonee, to hunt. I bargained with a company in Menomonee to take all the venison I could furnish up to the first of January.

As we started out one day for a hunt, we came to a place where twelve or fifteen elk had just been. We followed them up all day, but without success, and at night when we might have come within shot of some, a stream, the Red Cedar river, lay between us and we could not cross it. We were probably fifteen miles from camp; we, therefore, temporized a camp for the night, and in the morning retraced our steps, hunting as we went. When we came to the place where Jonathan and I had left the rest of the party the day before, we heard a man calling as if in great trouble. We went on until we saw one of our men down on a flat, running as fast as possible. We tried to get his attention; but he could not hear us for his own noise. When we did get to him, we learned that he could not tell where he had hitched his horses and he was frightened. He had shot a deer and could not find that either. His trouble was too great to keep it all to himself. After finding them we returned to camp, tired and hungry, and somewhat out of humor, because of our poor luck on our first elk hunt. We traveled through a rough, hilly, and heavily timbered country, and with comparatively poor success until after the first fall of snow, when I made good hunts through the season.

Deer hunts would become so monotonous and devoid of interest to the reader, if I were to attempt to relate a half of them, that I shall only speak of a certain few cases, letting the rest pass for just as good, but not, therefore, necessary to be described. Once when I had gone out with an insufficient supply of balls, I came very close to a large buck and shot it. He ran off a ways and then fell down, and, thinking to save my balls, I was going to hit him in the head with my hatchet to kill him. I knew he was badly wounded. When I was about six feet from him, he jumped up as if ready for a fight. His hair was all set forward, his tail stood erect, and the position of his head betokened his intentions. As I thought him ready to jump at me I jumped back; but in doing it hit my heels against a stick, which was partially concealed by the snow, and it was my turn to lie down. So down I went. I threw up my hands, intending to catch him by the horns if he persisted in displaying the aggressive. There was some nobility about the deer, as I have always claimed for the family, he did not propose to kick a man when he was down; but he stood there, meantime looking down at me, and I lay there looking up at him for several minutes. My eyes became his master, and after a little while I succeeded in crawling out of his reach; when I decided that I had balls enough, and I shot him dead.

Jonathan and one of the men one day found a bear in a hollow tree, and as it stuck its head out of the hole they shot it. It brought nearly thirty dollars. Later in the season some exceedingly severe storms made it so tedious to be out of doors that we went home. This winter was the hardest on deer of any I have ever known. White men and Indians slaughtered them in great numbers. They would put on snow-shoes, and taking a hatchet, but no gun, would strike them down. The snow was crusted, and would bear a man's weight, but the deer, falling through, would be so crippled in their traveling that they were easily caught. One man told me that he killed ten in one day, and that in some places the Indians had taken them by hundreds. They were very scarce the next year.

IN THE WOODS IN NORTH-WESTERN WISCONSIN.

D. D. Streeter, of Bernardstown, Massachusetts, went with me into several counties in the north-western part of Wisconsin to hunt deer. We went in the fall, hoping to find the deer in good condition and in abundance. We went onto Elk Creek, where I had previously hunted; but the deer were scarce, and the forests were fast filling up with a growth of underbrush. We therefore went on thirty-five miles further, to the Red Cedar river. Mr. Putnam was again with me. We pitched our tent and set our traps, ready for work and a good time generally. While Putnam and I were setting traps, crack, crack, went the gun at the camp, where we had left Streeter to settle our housekeeping arrangements. When we returned we found his booty was a big pile of prairie chickens and partridges, enough to last a good sized family for a week. Streeter set a trap for a deer on one of their trails close by the river, and in the morning we found it baited with a beaver. This was altogether new work for him, and he was delighted with his success. I went eight miles from camp onto Pine creek and trapped. My best day's work there was the catching of one otter, one mink, three beavers, and eight muskrats. There Mr. Putnam tried the old, but fatal plan of cutting down the dam to catch the beavers. He did let them out; but he caught only two from the four or five dams which he cut into. We stayed as long as we had intended to, and were well satisfied with the trip.

ABOUT EAU CLAIRE.

In the fall of 1858 Wm. H. Landon, Jonathan Cartwright and myself went to Eau Claire and the surrounding vicinity. We found the deer still very scarce, owing to the severity of the season two winters before. We therefore decided to make trapping our main business.

Having trapped for a time on Gilbert creek, we went onto Wilson creek. There I found a beaver dam which had on it the most new work of any dam I've ever seen. I went onto a hill to look down upon it, and it seemed as though the little fellows had chopped down a forest for the fun of seeing the trees fall. They lay in every direction, and being in keeping with the description of beaver cuttings already given, they would have stocked the cabinets of the country with the most valuable specimens of the sort. The dam measured six feet in height; the pond was therefore large, and there were a number of canals running into it. In my ignorance of the shyness, and, shall I say exclusiveness, or fastidiousness of the strange creatures, I frightened them away, and I failed to catch any, though I may except just the foot of one of them, for that I did catch. What is theirs belongs to them, and I was so ignorant of their habits that I trespassed upon their homes more than a skillful beaver hunter would dare to do. But I learned wisdom by my failures, and after a time I began to know and to catch them better.

We caught quite a number of otters, beavers, coons, and mink, while our camp was on Gilbert creek. One night we caught three otters and two mink. We were eight miles from Menomonee creek, and we had traps set sixteen miles south, on the Ogalle river. One day as I was going down the river I saw a coon's track, and knew that it was a new-made track. There was a sharp bend in the stream, and the water was at this point frozen about half way to the center. The water had fallen, the coon had gone into the stream, and sat near the bank with its head out of the water, but underneath the ice. I stepped onto the ice to get a good aim at my coon; but as I flung the stick from the edge of the ice, the ice broke, and I went head first into the cold stream.

One morning in December, after a snow storm, we thought we would go out and have some sport with some wild-cats. Two and a half miles from us there was a family of them hidden in some rocks. When about half way there I saw the tracks of seven elk, and I told the men I would follow them up, and they might go on. The elk had gone towards a mountain, and as I neared its top I saw one standing on a knoll not far from me, and as I fired my gun, another one nearer to me jumped up. I had not seen it; but I found that the ball had hit the mane, and having cut off quite a lock had glanced off, and the animal went on with the others unhurt. I tried to get in ahead of them, followed them up several miles, and then I saw one behind a fallen tree top. As I fired, it ran off. I had hit it, but too far back to kill it at once. I followed it up, and every few minutes would see where it had lain down. I traveled as fast as I could, and after a time, when I had reached a hill on the north branch of the Ogalle river, I saw elk again. I slid down the hill on my back to avoid attracting attention; but they were too much for me, and finding that I could not get a shot at them, I would shoot and I did shoot off my gun, and felt satisfied that I had had my own way about one thing.

The sun was then not more than an hour high and I was eighteen or twenty miles from camp, out in the cold traveling over deep snow with no hatchet or any matches, and in a strange country. I took my back track and followed it without trouble so long as daylight lasted. Just as that disappeared five deer started up near me, and one large buck was foolish enough to stand still and look at me: foolish, for I killed him. There was a full moon, but the clouds shut back a good share of the light. I had shot my last ball. There was no time for me to stop to think; but I did think as I tramped along. I wanted to strike an old road that had been traveled some years before. I was again left to do the next best thing, to keep upon my back track, if possible. After a time I did strike the road and followed it until I came to the mountain upon which I knew we had made our camp. I must turn from the road to strike it. This was a nice point to accomplish, especially in the dark. I knew that there were some very steep, rocky places, and knew that I must avoid them. A little to the east of our camp there was a windfall. I happened to get into it, and knew that I was not far from camp, but sweet as was that consolation, that windfall did not strike me in every respect with pleasure. I've heard of windfalls, and that they have been known to do good things for one; (this one did for me;) that they have been known to set people right up in the world. This one set me up in the world a good many times, for there was no way for me to get through it, but to get over a good share of it, for many of the fallen trees I could neither get around nor under. When I thought I must be close to the camp I called to the men and they brought out a light and I went in.

In the morning I went with Jonathan to get my game. Having found the deer and having hung it up to keep it from molestation, we went on to get my wounded elk. We found that a lynx had followed my track for several miles after I had left the deer the night before. As we neared the spot where I shot the elk I found where it had dragged its feet along, unable to keep them up any longer. When I saw it I told Jonathan to shoot it. I felt a little roguish just then. I knew it was dead ; but he did as I told him : he shot the dead creature and accepted the joke with good grace. We hunted elk the rest of the day, and at night camped out, and on the next day on our return to camp found nothing but one marten.

The day that Mr. Landon and Jonathan Cartwright started on their wild-cat chase they set some traps on the rocks which we called Wild-cat Rocks. The cave in these rocks, in which they were hidden, is three and a half feet high, five feet wide, and runs back straight for ten feet ; then turning to the right at right angles there's a passage fifteen feet long, two and a half feet high and three wide ; then it turns to the left and I do not know its dimensions. 'Twas a grand, good place for the ferocious felines. We had seen a very large cat there and had set a dog upon it ; the dog was badly whipped. One of the traps they set for the cats, and had fastened to it a dry pole about fourteen feet long. A cat had carried off the trap and the pole, and had gone into the cave. Mr. Landon made a torch of some white birch bark, and taking his rifle entered the cave. At the second bend he could see the cat ; he wounded it, and it jumped at him, and would have clinched him, but for the pole which caught in some rocks. He used his gun to ward it off, and he said that he backed out of the cave to the best possible advantage and as fast as he could. I believe the man told the truth. He succeeded in getting the cat out too, by pulling upon the pole whenever he could get a chance. After drawing it out he killed it. It was the largest wild-cat I have ever seen.

Soon after that, I left the men and went on twenty-two

miles and trapped on Mud creek, and caught several mink, foxes, wolves and coons. One day I caught seven coons. The men left the camp, and Jonathan Cartwright and Landon went onto Elk creek, thirty miles from our Gilbert creek camp. One night while trapping there he caught four otters, out of the five that came up the creek, and also two beavers. Two of his traps were carried off that night. Mr. Landon went onto Bloomer prairie after foxes. The last of February Mr. Putnam went with me onto O'Neal's creek. There we found fur very plenty. When the ice began to melt in the spring, by a series of severe exposures I became sick. Mr. Putnam went out to the settlement nearest us, to get provisions. He was to send them to me by Mr. Landon. The wind changed and "came from out the bad weather corner;" it snowed until one could scarcely see anything. Night closed over me; but no one came to my relief. My head ached, my bones ached, I was hot, was cold, and I was alone and lonesome. The night was a long and dreary one to me. When the storm ceased the snow lay twenty inches deep on the level. Early in the morning Mr. Landon reached me. In a few days I was decidedly better; but after staying two or three weeks and I was unable to hunt, I returned to my home. All things considered, we had a very fine time trapping and hunting. We caught fifty otters, forty beavers, one elk, one wolf, ten deer, eighteen martens, two fishers, thirteen wild-cats, ten foxes, three hundred muskrats, one hundred coons and one hundred mink. We caught two black coons. I have never seen any others like them. My partners were determined to make me like coon meat — steak, or roast, or fry, or in any shape they could cook it. They seasoned all of my food with the oil, but as I could not learn to like it, they almost starved me and I was therefore glad to get away from them.

IN EAU CLAIRE AND DALLAS COUNTIES.

After remaining at home for a while, I returned with Streeter and Putnam to our hunting grounds. We first went about sixty miles up the Chippewa river, then went to Vanville, about twenty miles from Elk creek. There we killed some deer. The snow was about two feet deep, and there was a good crust upon it. I thought I would go out and catch a deer and bring it in alive. I therefore equipped myself with ropes and strings, put on some snow shoes, and on finding some deer I followed one several miles. She could not endure it any longer, and turned about for a fight. She began to stamp down the snow to get a good standing place, and assuming the position of one previously described, she jumped at me. This time the deer kicked me on the side of my head, and I tumbled over, head down in the snow, and heels up and out of it. She ran on, and left me to help myself out of my trouble. Those of you who have ever worn a pair of snow shoes know how difficult it would be to get one's self back into standing position, with the feet well planted squarely underneath you, if you should chance to find yourself turned topsy-turvy, as I found myself just then. I did not mean to be joked and fooled in that way, and seizing all the determination I could scare up, I loosened my snow shoes and started after my deer creature. She had gone a few rods, and had another spot of ground already stamped down, and was ready to jump at me again. This time I escaped the fury of her foot, and I caught one of the hind legs and held it. She soon gave up, and fastening my ropes to her I drew her into camp. The next day I caught four more, and got them into camp alive. We built a pen for them of tamarack trees, and kept them through the winter.

In March we went into Dallas county. We stopped at a logging camp, and waited for the ice to break up. The boss of the camp was a Frenchman, who claimed to be very expert in managing a canoe. He wanted to show us how well he could handle one. He took mine and started out; but it was so much lighter than any he was accustomed to, that the first stroke he made with the setting pole slipped the boat out from under him, and sent him under the water. It was a very cold morning, and when he came up out of the water he took a bee line for the camp. He was well cheered by the spectators of the successful launch.

We went about five miles up the stream, and found deer in the greatest numbers I have ever seen them anywhere. It was here that Streeter and I watched to see how near a deer would come to us, as described under "deer hunting." We caught a good many animals, and when we had thinned out some of them in this locality, we started homewards. Twelve miles down the river we found our way blocked by logs. The water was high and swift. Some trees were turned over into the river, and many times it was difficult for us to find a way through the obstructions. In one place where some logs were bumping away at a fearful rate against a fallen tree, Streeter's boat, which was heavier than mine, found its way. He could not steer clear of them, and he came near to being upset. We trapped all along on our way home. This trip was the fourth one I had made in this section of the country with Mr. Putnam, the second with Mr. Streeter.

May we three meet again upon some fighting ground, where as before we shall be pitted, not against each other, but against the bears and the deer.

TRAPPING IN MINNESOTA.

In the fall of 1869 one of my neighbors, a Mr. Perry Sweet, went with me to trap in Minnesota. We drove a team, carried a tent and what was necessary for a good camping outfit. After crossing the Mississippi we camped near Postville. There we fell in company with a Mr. Fisher, who was also going to Minnesota. When we came to the Turkey river we found the water very high: the bridges were gone, and it was with some difficulty that we crossed the stream. Paul Cartwright, my son, who was then living a short distance beyond this river, went on with us, and a Mr. Ackley being added to the company, filled our ranks. We had a very hard time in traveling through this section of the country, because of some severe storms. Some nights we could not camp out, but were obliged to seek refuge, like fugitives in the days of underground railroads, in some barn or shed. Good roads were washed out; the low land was covered with water; mud was deep and bridges were gone; but had you tried to stop us, you would have found that we were gone too. We were living in high hopes of better days to come.

We stayed two nights at Clear Lake, and there the boys caught some ducks and geese, all that two men could carry. On our way to Coon Grove we had a sorry experience in crossing sloughs.

To any who have not enjoyed the luxury of being sloughed, it may be well to say that the low, wet land through which sluggish streams pass are called sloughs. These vary in width from a few to many rods. Many of them are covered with a thick sod that will bear up a team; others have a sod or reed bottom that will also bear up a load; but the mud sloughs, with neither top nor bottom protection, which must be crossed, furnish sufficient variation from the monotony of prairie travel. You force your horses to plunge into uncertain depths, and are fortunate if their feet are not fastened in the deep, unctious mud. If they are sloughed, your only remedy is to plunge in yourself, and keep their heads above water while you get them loose from the wagon. After they are safe, your remedy is with your heavy rope, carried for the purpose, to join horses and wagon together, giving the horses the benefit of solid land to stand on while they haul the wagon out. After the load has been carried out on your shoulders, unless the ladies prefer wading, they must be transported by the same method. By the time all this is accomplished, you have accumulated on your person and clothes an abundance of the thickest and richest soil imaginable, in addition to the inexpressible sensations of having been sloughed.

Coon Grove is situated upon a beautiful elevation, and it is surrounded by the best of muskrat marshes. At Jackson we procured our supplies for the remainder of the journey. We camped a few nights after leaving Jackson by a beautiful little lake, near which there were other smaller lakes. Whoever has lived in Wisconsin or Minnesota, though he has traveled but comparatively little, knows how beautiful the tiny clear lakes are; and those who will but look upon a good sized map will see there are a great many of them, and he may reasonably expect that there are many others which are not marked there. These states are thickly sprinkled over, like spice on a savory dish, with the tiny beauties. On this lake, by which we camped, we began our work of destruction. It was fun for us, but death to the rats: it made lively times for us, and for them while they lived. Paul and Ackley caught one hundred and thirty in one night. Whenever they began to get scarce, Fisher and myself looked up new trapping grounds.

Sweet was "chief cook and bottle washer," and we supposed we should have a sweet time eating the good things he would make for us. In honor of his high expectations, 'twas probably that, he christened the bread, biscuit, or food, or what shall I call it? with the name "marble cake." But, alas!

> "The best-laid schemes o' mice an' men
> Gang aft a-gley."

Though we will not do our friend the injustice to suspect him of roguery — for he had himself to eat of the same that he provided for the rest of us — we do think, if not very wise in choosing the proportions of his bread ingredients he could not have done but one thing wiser in the selection of the name for the —. Marble is hard, and so were the biscuits; marble is heavy, and so were they; marble is varied in its colorings, and so were they; streaked, and they were too. Marble is not found in the building and furnishing materials of every man's house; neither is the like of those biscuit upon their tables. Biscuit? Yes, that's the name of those things. He wasn't used to such, neither did he get used to the eating of them while he was the baker. He was sharing the fate of many a cook, losing the relish for his food, simply because he made it himself.

Mr. Fisher is a tall, heavy, energetic man, and he told the boys that he was sorry he could not have some man accompany him who could stand it to travel with him, that Cartwright was too old. One time on starting out we took a dozen of those biscuit along with us, and some salt, not that we might be able to keep them, but to salt what game we should cook, while away from the rest of the party. We started in a northwesterly direction, intending to strike Pipe Stone county, and if necessary to go on to Dakota. Mr. Fisher started out at a fearful rate, and I told him he might go on; for I could not travel so fast; but he stayed with me. At noon we came to a lake, where he wanted to stop and eat his dinner, I could eat all I wanted while traveling at my slow rate; he was to stop, as he would doubtless overtake me. We did not find any place at night where we dared to start a fire, lest we should set the whole prairie on fire. The next day I killed an elk, and I said to Fisher that we must go back to our home camp, and get the team to help us draw in the elk. He knew that he could not stand it to travel any further; but I thought he could; for he could travel a great deal further than I could, and I knew that I was able to go back to camp. I proposed that we take a shorter course, striking at once straight for the camp, and said that if we could reach the road before dark, I knew that I could shorten our distance by several miles. But he could not reach it. We had a sorry time of it traveling that night. Fisher knew I was lost, that it was not safe to follow my lead, that he was not able to keep up any longer, and he was determined to lie down upon the wet ground, though he could have no fire, and no protection. I used a little of a father's dictatorial style of argument upon him, as I thought it would become my age, and when we had reached camp he heard the first of the story, that it took so long to tell, about his having company equal to himself in traveling abilities.

The next morning it fell to my lot to go with one of the men after the elk. Mr. Fisher was not able to be up. We were gone two nights and fared rather hard in the time. The animal weighed about five hundred pounds. We moved our camp eight miles to the south of us and found good trapping grounds. We saw two large droves of elk : in one we counted seventy-three, in the other one hundred. The prairie had been burned over, and there being no trees to hide us from them, we could not approach within fire of them. We were in Nobles county, Minnesota. We were there successful in our trapping. We caught fourteen badgers, two otters, several beavers, sixteen skunks, eighteen foxes, twenty mink and twenty-seven hundred muskrats.

We found twelve families in Nobles county who spent the most of the time in trapping, but who were not very successful because they used so little skill. Rats were the principal game, and the skins were legal tender. While we were there a Sabbath-school was organized, and every man would give a certain number of rat skins towards procuring a library. If they bought law books with their money, I think the petitions therein contained must have been thoroughly rat-ified ; or had any of the buyers objected to such a selection, their objections were sufficiently ratified to secure a change. There we learned of a woman who was a practical hunter and trapper, and who was enviably successful.

I left my boat, traps, and tent, and we returned to our homes in December. It was cold and the return was tedious. Soon after reaching home Paul and I went back to Minnesota to trap again. We had severe storms to face : some of them were so severe as to prevent our traveling for a time. The snow was deep, and when we were beyond any habitations, and had our heavy packs upon our backs, we found the traveling very tedious. For the first time in my life, I gave out, and I gave my pack over to Paul. The next day, however, found me better, and ready for work. On our first day at trapping we killed one hundred and twenty-two rats, and twice that day we killed five at a shot. We caught four hundred in one slough, which was not a very large one. We thinned quite a number of sloughs of their rats, and after a very successful trip of a few weeks, for one of the sort, we left the grounds and returned to our homes.

IN THE COTTONWOOD COUNTRY.

In February, 1869, accompanied by Mr. Isaac Heritage and Mr. Darius Reed, I went into Minnesota to hunt along the Cottonwood river. After leaving Garden City, having hired some one to take us into the woods, we were overtaken by a storm and were obliged to stop on our way. The following day being clear, but cold, we resumed our journey, and, when in another day we neared Mound creek, our tribulations increased upon us in full measure. There came up very suddenly a violent wind storm which swept everything before it. It capsized our load, and our driver being thrown overboard, the horses started off as if to have things their own way; but, as men are apt to do when lost, they circled to the left, and soon came back. There was a light crust on the snow: the wind took it up and the loose snow from underneath it and at times almost blackened the air. The storm was, however, a very severe "white storm," such as raged with destructive fury in the winter of '72-3 in Minnesota and other parts of the West, the accounts of which were read by thousands of people, as many miles away. Our first hunting quarters were in a dug-out, such a place as many pioneers have not only heard of, but in which they have also lived. It was a ten-by-ten feet room, dug in a bank, high enough for us to stand in erect, was stoned up in front, and it had in it a fire-place. In that place we made ourselves quite comfortable. A Mr. Walker was our landlord, and furthermore he seemed to be the landlord of the country round; but in spite of his remonstrances against our hunting or trapping anywhere about him, we did carry off some valuable skins. I was again caught out in a terrible storm, and then became acquainted with a man who, having once been out in the midst of a white storm, had frozen his feet so badly that they came off at the instep. He could not reach any surgeon; he therefore had made a saw of some pieces of steel that he found in his house, and with it sawed off the bones in his own feet. His recovery was the reward he received for his courage.

We had very good success in trapping, and when after a few weeks we returned homewards, we started down the Cottonwood in two light boats, which we had with us. The stream is crooked and in many places very rapid, two conditions that, in conjunction with our style of craft, greatly enhanced the pleasure of a rough out-of-doors experience. We traveled by day down the river, and at night would pitch our tent in the most convenient place that we could find near by. After rowing our way through the Cottonwood, along the Minnesota river, and into the Mississippi, we reached Prarie du Chien, from which place we went to our homes by rail, having had no serious trouble with our light boats, except when going through Lake Pepin. We were in Minnesota at the time of an Indian insurrection, and saw many evidences of their vengeful wrath.

AN ENCOUNTER WITH AN EAGLE.

Early in the winter of '74-5, while hunting on the Yellow river about thirty-five miles from Chippewa, Wis., Mr. Van Antwerpt, who was then hunting with me, went out one day to get a deer that we had previously killed. We had hung the deer on a tree for safe keeping. He took it down, but had gone not more than forty rods from the tree, when an eagle, coming up from behind him, flew just over his head, and on, about twelve feet in front of him, then turned about and planted itself upon the ground in a decidedly combative attitude. It stood upon one foot, the other was uplifted as if ready for a fight; its head was erect, its sharp eyes as sharply fixed upon the man who was carrying off the deer, upon which it was probably about to feed; the feathers were all turned forwards, and "stood on end," and the wings were partially spread. The man threw down the deer, picked up a stick and throwing it at the eagle he struck the angry bird, but broke the stick. The eagle retreated about six feet, then turned about and faced him again, assuming the same fighting posture as before. The man picked up another stick and this time started for the bird, which at the same time advanced to meet him, and when each stood still they were within four feet of each other. He so wounded it that it could not fly, then taking another stick he started again for the bird, which was now walking off as best it could. Soon it turned upon him with an evident desire to fight it out as it had commenced; but the man soon killed it. It was a golden eagle, and measured two feet and ten inches from the bill to the tip of the tail, and seven and a half feet across the wings. It is now in a college cabinet, a fine specimen of its sort, and as an individual bird the reminder of a singular freak of our "historic bird."

VIII.

A TRIP TO LAKE SUPERIOR.

BY E. S. BAILEY.

(*Written August*, 1870.)

A few summers ago, through the kindness of President Whitford, the president of one of our Western colleges, I was invited to accompany him upon a geological and pleasure trip to Lake Superior. There were two others in our party, a school-mate, and last, but by no means least in importance in a trip of this kind, was our friend David Cartwright, a hunter and trapper, with whom the readers of these pages have already become somewhat acquainted.

We all desired to see some of the mines and mining towns of this famous iron and copper region. We all desired to experience the realities of camp life, to get outside of fences, and inside of the woods, where the bears and the bugbears might possibly meet us on our way, and we in turn might try our rifles and traps upon them. But the party was to a certain extent divided against itself, the first two named of said company desiring more specially to visit the mines, the other two preferring the depths above ground, to those beneath it. The details of this trip will not be fully recounted here, as we like best to talk to interested listeners, and our readers are supposed to be looking now, for a recital of the woodsman's sports, and not of the miner's home and daily toil.

Being amply supplied with blankets, clothing, and what is necessary for "roughing it," not to mention here a good stock of provisions, we left our homes in M——, to take the night train for Green Bay. We arrived at this place early in the morning, having had a quiet and safe journey. A short time after our arrival, we got on board the side-wheel steamer Gov. Dunlap, bound for Escanaba. The steamer soon started. Our course now lay near the middle, and the entire length of the bay. It is one hundred and twenty miles from Green Bay to Escanaba. The first part of the journey was very fine and we enjoyed it much. In the morning the bay was smooth; there was scarcely a ripple; but by noon a breeze sprung up, and at three o'clock, when we first sighted the light-house (off Escanaba), the wind was blowing a gale. This feature of the bay was by no means the most enjoyable, to us land lubbers. The scenery along the bay was very fine. We arrived at Escanaba at about four o'clock in the afternoon. The town presents an appearance by no means flattering at first sight, and a stay of two or three hours did not better the impression. There were several fine houses and one or two good hotels; but the yards were not fenced in, and even the pine stumps stood in the middle and on either side of the roads, and sidewalks: possibly they were left as an affecting reminder of some beloved pioneer of that immediate section of the country. The attractive feature of the place is its situation; it is just on a point that makes out into the bay, and is protected by high bluffs, which lie back of it. This place is always cool, because of its lake breeze. The iron-ore dock is the greatest curiosity there, but to describe it without a closer examination, I could not.

At six o'clock of the same day, we got on board the passenger train and started for — some place, — we did not know where. As soon as we left Escanaba we began to enter dense forests. For miles and miles on both sides of the track, there was nothing but trees, trees, trees. The train stopped at several stations, and took on passengers. These stations need but little comment; a log hut built near the track, and under the protection of some forest trees, being the only station-like object upon which to base a comment. We had been told that if we would watch closely, when we were about three miles out from Escanaba, we could see a camp of Indians. We looked, with all our eyes; but we did not see Mr. Lo, in his native laziness.

After one or two stops at these dreary stations, I raised the enquiry as to where we were to get out. I was answered that no one knew; but that we would find out before leaving the train. I thought at first this was intended for a joke, but afterwards learned that it was true.

The conversation of some of the passengers, who had doubtless had a residence far back in the woods for many years, was amusing. A single remark will suffice here. A father and mother had been with a sick boy, to see a doctor, and the mother in answer to the enquiry, "What is the matter with your boy?" said, "Why! bless you, mon, it 'pears as though his swallower is stopped up, and he can't drawer a bref." She said one of her boys had just died, but she wouldn't care any more for him if this one would live.

Just as the sun was setting, the train stopped and the baggage-master commenced to tumble out our baggage upon the ground. By this I knew when and where to get out. Three shanties and a hand-car house composed this village, and it was named Shaketown. Through the kindness of the section boss we obtained permission to take up quarters in the rear end of the car-house. After moving into our humble dwelling, I took an inventory of stock, viz: two trunks, a box of provisions, two guns, an axe and a long-handled frying-pan. To get supper was the next part of the day's doings. Hot tea was procured from one of the neighbors, and this with the provisions we had, made us a nice supper, and after partaking of the bountiful repast, we soon retired, not to downy beds such as they are supposed to be, but to hard beds down on the floor. But we rested. The next morning's sun rose bright, and with it the whole party of pleasure-seekers. An hour was spent in looking about, before we commenced to get our breakfast. Coffee was made in the four-quart pail, and the services of the frying-pan were brought into use in rendering some speckled trout, which the President had purchased, into a palatable condition. Having discussed the merits of our breakfast, we then discussed the propriety of officering the squad. President W—— was elected chaplain, captain, and commanding officer of the commissary department; Mr. C., the trapper, a builder of camps, and baggage-master; our friend Will R——, a dish-washer, and myself a cook and compounder of drinks.

Thus commissioned we started out about seven o'clock in the morning, to hunt deer, catch fish, or visit Smith's iron mines. By following a trail, we hoped to reach the latter place, in good shape in the course of two hours. To say nothing better for the plan, it embraced enough to keep us busy the two hours. Eighty rods out from the railroad track, we were out of all civilization, save such as we carried with us, in a dense, deep forest, wild as wildness itself. There was an indescribable grandeur in the feeling, that for once we were where none but the wild beasts roam.

After being out some two or three hours, we scared two

deer so fearfully that they left the bank of the lake, where they were feeding, and plunged off into the woods. We could have shot at them. Of course we could; but having only a few over five hundred rounds of ammunition, we did not care to waste any, and yet to be honest with you, we would have given it all if we could have captured the deer. A half mile further on we commenced fishing for trout. We waited long and patiently, but the fish, with more precaution than patience, waited longer. The fish did not take a bite that time; neither did we get a bite at them. We were quite discouraged, and almost disheartened over such luck, for according to the stories told us, we expected to get several deer, and a barrel or so of trout. No such luck was for us, so we started out to find the iron mines, feeling sure that, at least, these would not run away from us. We traveled until nearly two o'clock, and did not see a sign of the mines.

The fever for killing some game, so overcame Mr. C. that he left us to find the mines, and he was sure he would find the game. We lay down on the moss, under some stately pines and slept an hour away. A tramp of another hour brought us out at the head of the beautiful lake, which we afterwards heard called Blue Lake, a name significant of our feelings just then.

Here we first found substantial proofs that we were lost; and here first experienced the delightful sensation that flashes over one who suddenly finds himself literally and really lost. Tired! hungry! and lost! O most pleasurable topics for thought, to us who had little hope of rest, food or camp for the fast-coming night. But nothing would be gained by waiting at the head of the lake, so we started for the other end of it, where we chanced to see a saw-mill. Of all places mortals should take to walk in, this should be the last. I will not even except the laurel-brakes of our own country. Words fail to express to the stranger of such a region, the actual condition of the sidewalks of this city of trees. The shore was covered with debris, collected for many centuries. Fires had run through the woods and burned the most of the timber, so that almost all the young timber was down, and was not very nicely corded, I assure you. After climbing over this stuff, a huge pine stump would loom up in front of us, and this we would have to climb over or around. Trees twisted by the tornadoes of past decades had so scattered their branches that we were obliged to walk upon them, because we could not walk on the ground. After climbing over, and crawling under these a mile or so, we came to a thicket so dense it would almost shed rain. After having pushed through this, we had to wade through several rods of marsh, and then to get upon the trunk of some gigantic pine and walk upon that. Even this would sometimes prove treacherous, and let us through. Several rods of tamarack swamp now awaited us, and after we had walked, waded and in all kinds of ways tried to get through all this, we came to a spot of ground that lay high and dry—not a stick of timber upon it—and the whole nearly a rod square. Here we halted and shot off our guns, to prevent accident. Seeing two small birds, we shot at them, thinking it might be all we should have to eat that night; but both shots failed, and our supper vanished upon the wing.

 Very tired and lame with the labors of the day, we started on again, determined to push through or to fail in a glorious attempt. The solemnity of the passage through the swamp had not been broken by anything like laughter, for Will's declaration that this was no laughing matter, had toned us down, and had kept us suitably quiet. The crackling of underbrush startled us, for we had not given up the idea of seeing bears, wolves, and other such creatures; so W—— and I hastened onward to see what had happened to our commander—only to find that in jumping from one log to another, he had jumped beyond his length, and had landed in the mud and mire beneath him. At this we shouted in laughter. Why? Because we could not help it, and because he was laughing too. But we kept close to him, after this, to rescue him in case of future danger. Not traveling very slow, and greatly desirous of going still faster, he made another jump upon a huge birch log. The bark yielded to the pressure, and he, assuming the position of a bat lighting on a June bug, soon mingled with the underbrush beneath the log. Again we started forward, and after a mile or more of the same kind of traveling we came to the saw-mill. We were three miles from Shake-town. We learned also, that on the north side of Blue Lake there was a well-traveled road, and the distance was shorter by two miles. We had gone through a jungle that Africa can scarcely rival.

Three miles to camp! This we soon walked, singing hymns and whistling tunes, somewhat after the manner of the Israelites, when delivered from the Red Sea. We found our hunter at camp, with supper ready, but minus his game. Enough to keep a student three weeks suddenly disappeared. Next in order came the mending of torn clothes, bleeding hands and the repairing of the losses of the day. After our meager supper and limited exercises of the day, we laid us down to pleasant dreams.

Two days afterwards, we left this camp and went on to Negaunee, a distance of twenty miles from Shaketown.

Negaunee is situated at the junction of the Escanaba and the Marquette and Ontonagon railroads. It contains nearly twenty-five hundred inhabitants. They are composed of all nationalities. The greater portion of them have gone there to make as much money as possible in the least time possible. I think but few of them are rich. The business street is half a mile long and well built up on both sides. The buildings, however, are rudely built, and without regard to taste, or anything like architectural beauty or construction. On this street it is safe to say that every other store has connected with it a saloon. All the side streets were filled with saloons, and I believe that Negaunee has more saloons to the acre than any other place this side of the mining towns on the Pacific coast. It contains a single school with less than one hundred scholars. A machine shop and two blast furnaces are the chief manufacturing interests. The Jackson iron mine gives the principal life to the place. This mine is one of the largest in the iron regions and is situated about a quarter of a mile from the depot, and is, I should judge, about a mile in length.

While on our way to visit the Champion mines we passed through a place bearing an Indian name, which means in English, heaven. None of us had ever expected to pass through that place, (I mean Heaven) seated in a railroad car and have our neighbors playing "seven up" on the seat back of us; but one's expectations are seldom exactly realized.

We stayed a single night at Negaunee and there passed the night in camping out. We took our blankets and went into the woods, spread them and lay down to sleep. We were on a hill-side and I slept on the down-hill side, and when I awoke in the morning I found myself three or four feet from my chums. I think I've never slept better anywhere than I did that night. The sky served for the coverings, and the earth for the springs, while my boots served for a pillow.

We left Negaunee about one o'clock in the afternoon, and went on to Marquette, sixteen miles from the former place. The descent from Negaunee to Marquette is about seven hundred feet.

Under the direction of Mr. Cartwright we made our camp about a mile out of town, on the shore of the lake. Our tent was made of small pine trees. We commenced its building by placing a pole from one tree to another, and then put our little trees so that the branches were turned downwards, thus enabling them to shed rain readily. Having cut a good many boughs for the floor of our tent, after spreading them out and placing our blankets upon them, we soon had a splendid tent. We had not finished it when there came up a hard shower; but our tent sheltered us, and we were as "snug as a bug in a rug." This camp, just for luck, was called Bailey's camp. After spending three days of earthly bliss at this camp, we went on to Houghton. Before leaving, however, let me give you a description of the place.

Marquette is the largest of the towns in all the Lake Superior region. It received its name from the celebrated French missionary, Father Marquette. It is very pleasantly and picturesquely situated on a beautiful bay, which travelers say bears a very marked resemblance to the Bay of Naples. The town is built upon the circling shore, which rises in natural bluffs, street above street. A few years ago it was nearly destroyed by fire; it was in the winter, and during a severe gale. It was speedily rebuilt, and when I saw it presented a fine appearance. The bracing air at Marquette is its principal attraction for those seeking recreation. All the time we were there the weather was delightful, not too warm, as it was comfortable for one to wear flannel, even though it was August. This place is becoming more and more frequented by pleasure seekers. One of their rustic poets writes as follows about Marquette. (I copied it while there:)

> "The air is bracing, sweet and pure,
> Those who're sick can well endure
> To breathe it freely; while they stay,
> A balm they'll find it every day.
> We've mossy hills and shady dells,
> And rippling streams, where music swells,
> And pleasant sights the country 'round,
> As any in the realm are found."

The inhabitants and tourists pass a considerable portion of their time fishing and boating. There are few good drives, except along the shore. Some days travelers have splendid luck at fishing; but they happened to be out of town, and couldn't tell us how it was done. We were always out on unlucky days, only once, when one of the party caught a trout that weighed nine pounds.

To perfect this place for a summer resort would require only good bathing facilities. The water changes only six degrees during the entire year. It is rather cool for bathing, yet we could not resist the temptation to plunge into the pure, clear waters of that great lake. The water is so pure that, when calm, objects can be seen at a depth of seventy-five feet. There are no shells to be found on the shores of Lake Superior.

The local scenes are very fine, and some are very curious. As active business-like men are seen in its streets as in any other place.

Indians, with their squaws and papooses, are ever to be seen lounging around, or selling some of their trinkets. The windows in the business part of town are full of curious specimens of iron and copper ore, deer's heads with large antlers. Wild cats and panthers nicely stuffed are used as signs, and all sorts of Indian curiosities, from a full sized birch bark canoe to a very small pair of moccasins. The iron money, not coined, but in the form of bank bills and issued by the iron companies, made payable by their agents in Detroit or Cleveland, is common here. It is hard looking money. The first time I got any I thought I had been swindled to the amount of the bill.

Another curiosity is the construction of the wharves for the loading of vessels with ore. It is very much like the one at Escanaba, and, as I visited this one, a single description will answer for both. I should judge they were thirty feet above the lake, and extend eight hundred or a thousand feet into it. Upon this a train of thirty or forty cars loaded with ore is run; a trap door in the bottom of each car is opened, and the ore is conducted in sluices down to the side dock below, at which vessels lay. Thirty of these trains arrive daily with their loading of ore.

About six o'clock Thursday afternoon we broke up Bailey's camp, and started with our luggage for the steamboat wharf. At ten o'clock the City of Toledo left the wharf, bound for Houghton. It was a dark and stormy night, and we were glad enough to have the shelter of the steamer, for I think the romance of Bailey's camp would have been washed away by the drenching rain.

An Indian half-breed made sport for us an hour or so by giving us exhibitions of clog dancing, and other ways of tripping the "light fantastic." After he had danced out, we turned in for the night. We were all up early in the morning to see the sights. The steamer was just leaving the lake, and was entering Portage Lake, and would be at her dock in an hour and a half.

The scenery all along this lake is grand, and I cannot pass it by without a word. On either side of the lake, or rather the outlet of the lake, very high bluffs rise gradually from the water's edge, and a full mile back reach their summits. They are covered with a great variety of trees, so scattered that one cannot help admiring them. Here and there barren rocks loom up in bold relief. The scene was constantly changing, but was still as fine in one place as another.

This country was just as it was made by the Creator, undisturbed by man. We saw along the bank of the lake several Indian wigwams, or huts. They were rudely built, and only calculated for one season, and some for only a day. One camp I noticed in particular. It was made as follows: two stakes were stuck in the ground; across them was placed a ridge pole, some four feet from the ground; then two sticks were placed with one end on the ridge pole, and the other on the ground. These poles were then covered with birch bark. Under this rude covering there were eight Indians. There were two large birch bark canoes drawn up on the shore. My love for the Indian race was not at all increased by seeing this poor, lazy set of fellows. They were rudely clad, and eked out a miserable existence by hunting and fishing.

About seven o'clock we stopped at the wharf at Houghton. The place was named after Dr. Douglas Houghton, one of the very earliest settlers in the northern regions. He was afterwards drowned in Portage Lake.

As we could not find a good place to camp, we secured a house with ten rooms, a store front and a large wood shed. We thought this would accommodate us for room. We very soon made the acquaintance of the people in the adjoining house, and from them received many favors during our stay.

Across the river, but a little higher up the stream, is a little place called Handcock. It is about the size of Houghton, and about as dead; but this place has the stamps, the ponderous iron stamps, I mean, which are used in crushing the ore to secure the copper. We all crossed on the ferry, intending to see the sights about the mines and stamps there, but we were told that the best mines to visit were thirteen miles north. President W. and myself determined to go on to see them. Mr. C. and Will were to look at the sights in the immediate vicinity of Houghton and Handcock. We hired a horse and buggy and rode over hills and mountains, and through some timber, where the trees were so tall that we had to look twice to see the tops of them.

Our visit to the Hecla and Calumet mines was sufficiently interesting to us to make us forget our dinner. We afterwards visited the Quincy mines and some others, and all of the visits resulted in united pleasure and profit.

On returning to Marquette with the rest of the party we learned that there was to be an excursion to the pictured rocks. The President and myself went with the excursionists on board the "City of Toledo," which was to take us out that day. No pleasure seekers could be better pleased with the sights of a single day than were we on that trip. Our friend Mr. C. and his faithful companion did not see the pictured rocks, for there was pictured so unmistakably and indelibly upon their minds a hunter's return to camp, laden with choice booty, that they could not be persuaded to spoil the scene by leaving themselves out of the picture. If you ask if the picture is in the possession of any of the party, and if framed or otherwise on exhibition to inquiring friends; and if the deer, bears, wolves, and the trout are all there at the feet of our heroes, we have but to refer you to the reports of the old and the young hunter, as they gave them to us on our return to Shaketown, whither they had gone; or we will leave you with them to hear for yourselves what they have to say of the events of the day, and will only ask that when you have learned what you can of them, you shall sometime tell us whether theirs was the better choice. Do you leave it to us to tell you, we may say that after they came on board the train at Shaketown, and we found ourselves homeward-bound, we interested them for a while with accounts of the beautiful scenery which the famous rocks had furnished us that day, and of which you already know, or may with little effort learn. In return they told us how *near* Mr. C. had come to killing a deer; what fun the inhabitants of Shaketown had in killing a skunk, and the "Pick him up on a shovel, Mike," "Catch him by the tail, Johnnie." They told us how Will had been frightened within an inch of his life by a wolf, as he was coming into camp at night; how it was that Mr. C. was close by and there wasn't any wolf there.

The provision box they filled with whortleberries, twenty-four quarts, which they had picked in about three hours.

Our trip down the bay was without excitement. We took the same boat as on the upward trip. At Green Bay we looked about some. It is one of the oldest towns in the Northwest, having been settled about two hundred years. The appearance of the place also justifies one in this bit of history. A few hours by rail found us in our homes. I count the trip a profitable one, and think it did as much for us, as the adage of early rising promises to do for one. Our out-door life had given us increased health; our sight-seeing had made us wiser, and they both had given us a measure of that wealth that falls to a sound body, and an appreciative, happy heart.

This is the short story of a long trip. During this trip Mr. Cartwright became to us "Uncle David." By his observing the tops of certain trees we were always sure of the points of compass. By his careful and constant watchfulness he was ever pointing out to us the habits of the animals that we started up as we journeyed through the forests. The study of nature amounted to enthusiasm with him. He knew all about trees, brooks, and rocks. In the finding out of the secrets of the lesser animals, and by skill and cunning to arrange for their capture, was his delight. He carried his trusty rifle in one hand, a walking stick in the other, and with eyes bent on the ground he eyed every track and broken limb for some indications of approaching game. To see a place along a stream or by a lake where fur-bearing animals had been was to set up a land-mark for a future visit during the trapping season. He was kind and temperate in everything except in long journeys. He has since become so familiar with Lake Superior woods as to offer himself as a guide for visiting parties, and has given us all invitations to hunt with him, but it has been impossible to accept them. Should any of our readers be fortunate enough to have his invitation for a trip in the woods, or for a stay of a week or two at his camp, it would be simply the prelude to a pleasant trial of camp life and the furnishing of good things to be remembered in after days.

IX.

TRAPPING IN THE LAKE SUPERIOR REGIONS.

Mr. C. P. Clemens and myself commenced hunting in the Lake Superior regions in the fall of 1870. Mr. Clemens, whom we all called "Uncle," is a genial, good natured, happy old soul, who likes to be happy, and to see others so, even if he is obliged to confer the happiness himself. He is, therefore, both rich and benevolent, possessing and willing to impart, what so few possess, or know how to bestow upon others. He is a man of excellent principles, is strictly temperate, and never swears; nor does he use vile, indecent language, as some people suppose he must do, if he is a hunter. There is no occupation that obliges a man to swear, or that makes it necessary for him to become a drinker. It is a matter of grief to those of us who would follow the woodsman's craft, that we are set down as roughs, as profane, intemperate men, and yet we know that such opinions are born either of ignorance of the craft itself, or based upon the shiftlessness of those who follow the business in their own naturally shiftless ways. Uncle Clemens is a practical woodsman, but one whom we know can enjoy to an enviable extent a social gathering within doors as well as beyond the fences. Being an experienced hunter, I found him a good partner, and remember with pleasure those seasons in which I have traveled with my "uncle."

The country through the northern section of our trapping ground was specially dreary. North and west from

Shaketown the land was covered for miles and miles with burnt timber. Upon the high grounds south of this burnt district we found maple, birch, and hemlock trees. Upon the low lands, over which we traveled the most, there were cedar, spruce and tamarack swamps.

Our line of traps was so set that it required a march of a hundred miles to reach them all. We had nine camps upon the line. One of them, which was a logging camp, was our home run, and was about nine miles from Shaketown. There we kept our supplies and met in council, we two "good Indians" of the woodsman's tribe. We each trapped and camped alone the most of the time. My own camp was twenty miles from the home camp. Whenever we left one we would leave in writing upon the wall a statement of our successes and our plans until such times as we should arrange to meet again. We caught seventy-three beavers, fourteen otters, sixty-seven mink, ten martens, eight fishers, six lynx, four foxes and two hundred muskrats.

In the spring Paul Cartwright, my son, was with us. One day on reaching the home camp, when Paul's camp was fifteen miles west of it, and my own twenty miles southwest, I found that he was intending to come to my camp on a certain day, if it did not rain. But it rained. I waited several days, and as he did not come I went to the home camp and there learned that his plan was to start the day before. He should have reached me at night. I retraced my steps; but not being able to get through, and there being two camps between that one and mine, I stayed all night at the first camp. I there expected to meet Uncle Clemens, but did not. In the morning I went on to the second camp and there found the dog that Paul always kept with him, and one that could hardly be induced to leave him. He was alone, in a very sorry condition; he looked sad and forsaken; his head was shot full of hedge-hog quills, and he looked as though he had had nothing to eat for several days. This so alarmed me that I started at once for my own camp to find my son, knowing that if I did not find him there, or possibly somewhere on my road, I should have scarcely any hope of finding him. As I started from the camp I had a pack of seventy pounds upon my back. I had five miles to go over a rough road, or more literally, over our trail. As I went along, I shouted frequently, thinking I might possibly get a response, and reached the camp in an hour and a quarter. There I found my son. I had not thought of my pack until I saw him, when my nerves and my muscles so soon relaxed that I was unable to take another step without help.

After I had been home and sold my furs I went into the woods again, and commenced trouting. We sold nine hundred pounds, for which we received forty cents a pound. As soon as deer were in good condition we hunted them again and caught sixty-seven, for which we got from ten to twelve and a half cents per pound.

During that summer as I went one day to salt a deer lick, which was twelve miles from Shaketown, and as I was crossing a swamp on a corduroy bridge, I looked into the swamp and saw a bear coming onto the road. I shot him and he ran towards me, growling at every step. I set my dog upon him, but as they met, each one turned to the right and they passed each other. 'Twas rather ceremonious for the circumstances, however, for no sooner had the dog passed him than he wheeled about and pouncing upon the bear, clinched him. The bear tried to catch the dog, but the dog escaped and the bear made another lunge towards me. The dog caught him a second time, and he fell and died. I found that I had shot him through the heart. The bear probably had no idea of touching me. I think he did not see me; but he happened to run towards me, and I think because the road was better than in the opposite direction.

I returned to my home in the fall and remained until spring, when Uncle Clemens and I hunted together again upon our old hunting ground. During the summer and fall we caught ninety-seven deer, eight bears, about thirty-five beavers and as many mink. The mink skins were that fall unusually handsome. We sold six hundred pounds of speckled trout, and I made several trips with fishing parties. Paul Cartwright was with us five weeks during the time. Besides our deer skins, we sold nine hundred and seventy-six dollars worth of other stuff. We were out from June to January.

Early in July of the same season Willis P. Clarke, and Postmaster Paul Green of Milton went up to our trapping and hunting grounds to test the experiences of life in the wild woods. Very early in the morning of the first day after their arrival Paul Green went out with me to catch, if possible, some deer. I soon shot one and putting it upon my back we started for camp; but before we had retraced many of our steps he killed one and we went back into camp before breakfast with our two deer, and all counted it a successful trip for our sportsman. Our friend Paul afterwards killed four others, and our friend Willis also killed his first deer while there.

"Uncle Clemens" and myself afterwards went out from our shanty, accompanied by the two sportsmen just mentioned, to an old lumberman's camp about four miles from the mine. We arrived there about noon, and after dinner Clemens went down the Escanaba river to the "Cataract" for trout. Paul and Willis went with me to a deer lick, when on arriving we found that a bear had been poaching there. Leaving them to watch for the bear, which I believed would return before night, I went on to watch another lick. Their post was upon a granite ridge overlooking a muddy slough. About four o'clock, on looking to the north, the men saw their ursine friend approaching them. He came steadily on, crossed the slough on the fallen timber, and came directly in front of them. Although it was the first time either of them had seen a bear in his native wilds they took steady aim, and at the word both fired. The former put a ball through his back and the latter a charge of buck shot into his body. This unexpected reception rather disconcerted his bearship, and before he could recover his presence of mind another ball from Paul's rifle laid him out. He was a black bear, and in good condition would have weighed four hundred pounds. I returned to them just at night, having killed while gone two bucks. We then returned to camp and feasted on venison and broiled trout. When they returned to their homes they carried with them good cheer, and a high appreciation of the pleasures of such a trip.

During the following summer, or the summer of 1873, I went out with two sporting parties. The first party consisted of Smith, Baggs, Millard, and Stephens. The last one named came from Smith's mines. He is a man who is prompt in his habits and quick in his motions. He is well educated, and having traveled extensively, is trained in that knowledge that comes of a practical study of various peoples and their varied homes. The others came from Chicago. Baggs and Smith were in office employ of the Chicago, Burlington & Quincy Railroad. Millard was a clerk in a book-store.

As I knew them, Smith reminded me of Uncle Clemens in his kind and genial disposition, and true to the disposition which he represented, he was capital good company. Baggs was the joker, and kept us all in time and tune. Millard was the most persistent fisher I have ever known. Sometimes when he displayed his excessive persistency in quietly holding his fish pole, and when the water was clear and the sun was shining bright and we could see through it for a radius of several rods and there was not a fish to be seen, he evoked the time-worn adulation, and incentive to noble deeds, " Perseverance conquers all things," or sometimes when mirth got the uppermost of our sobriety in contemplation of the stern realities of actual business life, the great demands it makes upon our time and patience, some one was always ready to pat him on the shoulders and sing in his ear, " If at first you don't succeed, try, try again."

We went onto the Escanaba river and camped two nights a few miles above Smith's mines. When we were returning to our camp, which was a few rods from the mines, Smith in his jollity, jumped about a little too much for a man in a row-boat, and when he found the boat rocking too much he thought to sit down on the edge of the boat and behave himself, as I suppose ; but he found himself landed in the water, and for a time he did behave himself like a man, but a sick one, I mean ; for he took cold and was sick.

On our return to the mines the men wanted to go out to catch some deer. Paul and Jonathan Cartwright went with Baggs and Millard, and I went with Smith. I cautioned the boys not to get excited if they should see some deer, and told them when and how to shout. They were capital boys to mind ; they did keep cool, so cool and so unexcited that when a deer came within fire of them they did not offer to shoot it, or even to frighten the poor thing. I've never heard that they made much money on the sale of venison from that day's hunting. I took Smith to a deer lick, and left him to watch it while I went on to another one. On nearing it I shot a very large buck. It ran off. When Smith came into camp he said that I had killed a deer. I doubted it; for, as I told him, I had found a drop of blood on a leaf which was so high that if it had dropped from the wounded deer, as I knew that it did, it was shot so high on the body as not to kill it very soon. I had, therefore, not attempted to find it; but in the morning I did find it dead.

We went to Escanaba by the river. Our ride on the river took us sixty or seventy miles through a country in its native wildness. There was not a house to be seen along the river; heavy timber comes close to the banks. In many places the stream is very rapid, and there are occasionally perpendicular falls of water of six or eight feet. In many of the places through the rapids it is unsafe for boating; for there are large rocks on the river bed. When we could see our way out of these places, the men would begin to shorten their faces, and to wake the woods with their shouts. They had gone for fun, and for a thorough relaxation from the close confinement of business.

We had a first rate time fishing. We were six days going down the river. One night, just after sundown, we saw a deer crossing the river and coming toward us. Baggs and Smith both shot at it; but the deer made good its escape.

Stephens and myself stayed one night at Escanaba, then returned by rail, in company with another party which I had met there by agreement. This company was also from Chicago. Three of them, Howlan, Evans and Alexander, were conductors on the Chicago, Burlington and Quincy Railroad. Nichols was a news agent.

We went to Smith's mines, and stayed one night in our tent, about forty rods from the entrance to the mine, and the next day we went to the camping ground of the first party, above the mines. There we divided into two companies and spent the day in trouting. Three of the men went up to the cataract, and were not specially successful in catching fish. Two of the men went with me onto the Little North, a branch of the river. It is a small stream, but rapid. The timber is very thick along this stream, and a great deal has been blown down and into the water, making it doubly difficult to travel upon it. The scenery is very wild, and to many it might seem dreary. My Chicago friends thought they had never seen a wilder country than when upon that trip. We caught about one hundred and fifty trout, and returning to camp met our other friends.

The day following this they also proposed to have a change on the programme; so we started out to catch some deer. I went off and shot a very large deer. I took the saddles, which weighed seventy pounds, and returned to camp. As I neared the camp Alexander came to carry my load for me. He was like the little girl who thought "That's a good many for three;" seventy pounds weighed more than he thought it did. The next morning he helped me carry the remainder of it to camp. The head and horns he appropriated as trophies of his chase.

Our return trip to Escanaba was also down the river by boat. We had two boats, and as we neared the rapids it was fun for the men to watch each other's boats. I always stood when managing a boat, and I could not keep my equilibrium as we bumped over the stony bottom. Once, to the amusement of the men, I was thrown forwards, and lay across some of the men, with my hands and my head in the water. Another time I was thrown entirely out. Whenever anything of this kind would happen, our railroad conductors would call out, "All aboard!" Howlan and Alexander were large, two hundred pounders, and they made the woods ring with their invitations to a free ride and no accidents on this line. Whatever made fun for one party called the attention of the other, and no sooner would their eyes be off from their own course than they would bump against a rock, and the tables were turned: the laughers became the laughees. Frequently when a boat would be brought up against a rock, some of the men would get out, (the water was very shallow,) and pulling a rope at the bow of the boat, would pull it over. They were very kind to me. These two parties from Chicago were the only ones I've ever had offer to do such things for me. I thank them for their kindness; for the pleasure they furnished me I hold them in kind regard.

A little above Flat Rock we camped for our last night. There we found large piles of flood-wood, pine and cedar. The boys, for such you know we frequently call men, had a big bon-fire. I think 'twas large; for I presume they frequently had eight or ten cords of wood burning at once. Around this fire the boys went in for fun, and nothing but fun. Evans we called "Big Indian Me." During the war he was on the frontiers, and had seen a great many Indians in their own country. He had learned their war dances, and other of their performances, and he filled the tree tops with his war whoops, and the yells of the scalp dance. We had scalped no Indians; but we had, if you please, scalped other creatures in their native wilds, and we had the fire and the dance about it, in good imitation of the red man's wild exultation, "Me killed Cheyenne! me killed Cheyenne!"

I had been out with portions of both of these companies the previous year, and I remember the two seasons with pleasure. The returns of our bulleting campaign for the season in the electorial district above described, were eight bear skins, eighty beavers, seven otters, three fishers, thirty-five mink, over eighty deer, besides other cheaper furs. We also caught about six hundred pounds of trout.

During the summer of 1874 I again went out with pleasure seekers. The pleasures of the trip, fortunes and misfortunes, bear a marked resemblance to those of the previous season; but one of the men, Mr. Davis Rogers, either because of my own dexterity in deer stalking, or his want of it, resulting from inexperience, was sadly overcome with an idea he had of the difference between a professional deer catcher and a green horn hunter.

www.ingramcontent.com/pod-product-compliance
Lightning Source LLC
Chambersburg PA
CBHW021207230426
43667CB00006B/595